OECD PROCEEDINGS

Fertilizers as a Source of Cadmium

PUBLISHER'S NOTE
The following texts are published in their original form to permit faster distribution at a lower cost.
The views expressed are those of the authors,
and do not necessarily reflect those of the Organisation or of its Member countries.

ORGANISATION FOR ECONOMIC CO-OPERATION AND DEVELOPMENT

ORGANISATION FOR ECONOMIC CO-OPERATION AND DEVELOPMENT

Pursuant to Article 1 of the Convention signed in Paris on 14th December 1960, and which came into force on 30th September 1961, the Organisation for Economic Co-operation and Development (OECD) shall promote policies designed:

- to achieve the highest sustainable economic growth and employment and a rising standard of living in Member countries, while maintaining financial stability, and thus to contribute to the development of the world economy;
- to contribute to sound economic expansion in Member as well as non-member countries in the process of economic development; and
- to contribute to the expansion of world trade on a multilateral, non-discriminatory basis in accordance with international obligations.

The original Member countries of the OECD are Austria, Belgium, Canada, Denmark, France, Germany, Greece, Iceland, Ireland, Italy, Luxembourg, the Netherlands, Norway, Portugal, Spain, Sweden, Switzerland, Turkey, the United Kingdom and the United States. The following countries became Members subsequently through accession at the dates indicated hereafter: Japan (28th April 1964), Finland (28th January 1969), Australia (7th June 1971), New Zealand (29th May 1973), Mexico (18th May 1994), the Czech Republic (21st December 1995) and Hungary (7th May 1996). The Commission of the European Communities takes part in the work of the OECD (Article 13 of the OECD Convention).

© OECD 1996
Applications for permission to reproduce or translate all or part of this publication should be made to:
Head of Publications Service, OECD
2, rue André-Pascal, 75775 PARIS CEDEX 16, France.

Foreword

This publication contains the papers presented by representatives of governments and industry, and other experts in the field, at the **OECD Cadmium Workshop** held in Saltsjöbaden, Sweden, on 16-20 October 1995. The Cadmium Workshop was co-sponsored by the Swedish National Chemicals Inspectorate (KEMI) and the Dutch Ministry of Housing, Spatial Planning and Environment.

The Cadmium Workshop consisted of an opening plenary session, followed by two subsidiary workshops: the **Sources Workshop**, which addressed all sources of cadmium inputs to the environment (with the exception of fertilizers); and the **Fertilizer Workshop,** which specifically addressed phosphate fertilizers as a source of cadmium inputs to agricultural soil.

The papers published here were given during the Fertilizer Workshop. Its four "parallel sessions" dealt, respectively, with measures and techniques to reduce the cadmium content of fertilizers; implications of measures to reduce the levels of cadmium in fertilizers; accumulation in agricultural soils, and cadmium content in food and human uptake; and uptake into crops and bioavailability. Also included are the final reports of each session.

The papers given during the plenary session at the beginning of the Cadmium Workshop have been published in another volume in this OECD series, *Sources of Cadmium in the Environment.* They concern sources and pathways of cadmium in the environment; the fate of cadmium in the environment; environmental concentrations and trends; transboundary pollution and bioavailability; routes of human exposure and trends; exposure of ecosystems and trends; cadmium in waste; and trade aspects. **The companion volume also contains the papers presented during the six parallel sessions of the Sources Workshop** (which addressed, respectively, natural and anthropogenic sources; sources of inputs to the environment; products containing cadmium; various non-product sources of cadmium; other sources of cadmium; and cadmium in waste). The final reports of each of these sessions are included as well.

More than 150 officials and experts attended the Cadmium Workshop, representing OECD countries, the OECD's Business and Industry Advisory Committee (BIAC), and several phosphate-producing countries which do not belong to the OECD.

The Cadmium Workshop was followed on 20-21 October by a meeting of the OECD's Working Group on Cadmium Risk Reduction, which included policy experts from Member governments and representatives from BIAC. The objective of the meeting was to examine the reports from the sessions of the Cadmium Workshop and identify points that needed consideration by the Joint Meeting of the OECD Chemicals Group and Management Committee.

There was a consensus in the Working Group on the need to continue efforts to reduce risk from exposure to cadmium; however there was no consensus that *direct concerted* risk reduction action is warranted on an OECD-wide basis. The Group did agree that there were opportunities for Member countries to reduce risk by establishing or enhancing national cadmium risk management strategies. In addition, the Working Group felt that these efforts could be supported, possibly in the context of the OECD, with regard to activities aimed at collecting and sharing more information.

The Joint Meeting, on 8th February 1996, took note of the results of the Cadmium Workshop and Working Group meeting. It accepted the consensus view of the Working Group with regard to direct concerted action, and agreed to the recommendations concerning further work within the OECD on the collection and sharing of information.

The Joint Meeting subsequently recommended that this publication be derestricted. It is published on the responsibility of the Secretary-General of the OECD.

Background

In 1990, the Council of the OECD adopted a Decision-Recommendation on the Co-operative Investigation and Risk Reduction of Existing Chemicals [C(90)163/Final]. This OECD Council Act is aimed at the reduction of risks from chemicals to the environment, and/or to the health of the general public or workers. It is based on the premise that international co-operation in risk reduction activities can enhance the technical and institutional aspects of risk management in Member countries through burden-sharing and a reduction of duplicative efforts. Furthermore, such activities can lead to more effective use of the knowledge of risks being generated through, for example, national chemicals reviews and assessments; the OECD co-operative investigation of existing chemicals; and the work of other international organisations conducting hazard and risk evaluations, such as the United Nations' International Programme on Chemical Safety (IPCS).

OECD countries chose cadmium as one of five chemicals (or groups of chemicals) to be included in an initial pilot project on co-operative risk reduction. In 1994, the OECD published a "Risk Reduction Monograph" on cadmium,[1] in which its commercial and environmental life cycle is described and information is provided on international and national positions concerning cadmium's risk to man and the environment, as well as on measures taken by OECD countries to reduce such risk.

---

[1] *Risk Reduction Monograph No. 5: Cadmium. Background and National Experience with Reducing Risk* [OCDE/GD(94)97]. Risk Reduction Monographs have also been published on lead, mercury, methylene chloride and brominated flame retardants. **These and other technical reports prepared by the Environmental Health and Safety Division, as well as copies of relevant OECD Council Acts, are available at no charge from the OECD Environment Directorate, Environmental Health and Safety Division, 2 rue André-Pascal, 75775 Paris Cedex 16, France. Fax: (33-1) 45.24.16.75. E-mail: ehscont@oecd.org. For more information, including the full texts of all five of the Risk Reduction Monographs, consult the OECD's World Wide Web site: http://www.oecd/ehs/.**

Following publication of the Risk Reduction Monograph, OECD countries determined that more information was needed before they could consider whether OECD-wide cadmium risk reduction measures might be necessary. The Governments of Sweden and the Netherlands therefore agreed to co-sponsor a technical workshop which would attempt to fill any existing data gaps and draw together the necessary information on which to base a decision concerning further action.

During preparations for this workshop, a questionnaire was distributed which asked for country-specific information beyond that found in the Risk Reduction Monograph. The Dutch Ministry of Housing, Spatial Planning and Environment commissioned a detailed study of cadmium's sources and pathways, using data generated by the questionnaire and other sources. KEMI also commissioned a report (based in part on the OECD questionnaire) on phosphate fertilizers as a source of cadmium. The production of these reports, which have been published by the two countries,[2] laid the groundwork for the OECD Cadmium Workshop held in Saltsjöbaden.

This publication was produced within the framework of the Inter-Organization Programme for the Sound Management of Chemicals (IOMC).

---

[2] *Cadmium, Some Aspects of Risk Reduction* is available from the Dutch Ministry of Housing, Spatial Planning and Environment. It was produced by Jonathan Pearce. *Cadmium in Fertilizers* is available from the Swedish National Chemicals Inspectorate (KEMI). It was produced by the European Environmental Research Group Inc. (Lars Landner, Jens Folke, Mona Olsson Öberg, Helen Mikaelsson and Marianne Aringberg-Laanatza). Both were published in September 1995.

The Inter-Organization Programme for the Sound Management of Chemicals (IOMC) was established in 1995 by UNEP, ILO, FAO, WHO, UNIDO and the OECD (the Participating Organizations), following recommendations made by the 1992 UN Conference on Environment and Development to strengthen co-operation and increase international co-ordination in the field of chemical safety. The purpose of the IOMC is to promote co-ordination of the policies and activities pursued by the Participating Organizations, jointly or separately, to achieve the sound management of chemicals in relation to human health and the environment.

Table of Contents

Session A

Measures and Techniques to Reduce the Cadmium Content of Fertilizers

Issue Papers:

Managing the cadmium content of phosphate rock: a contribution to environmental impact mitigation
Arafat Ghosheh, Saleh Bashir and Lana Dabbas 13

Studies and research on processes for the elimination of cadmium from phosphoric acid
A. Davister 21

Cadmium removal from phosacid
R. M. Vermeul 31

Promoting the development and semi-industrial application of a potentially high performing process for cadmium removal from phosphate rock
Abdelaâli Kossir and Abdellah Chik 41

Report of Session A 45

Session B

Implications of Measures to Reduce the Levels of Cadmium in Fertilizers

Issue Papers:

The existing instruments for environmental, technical and financial co-operation regarding Africa and the Middle East
Marianne Laanatza 51

The importance of the phosphate sector to the economy of Senegal
I. Kotlarevsky and D. Fam 56

Agronomic implications of restricting cadmium content of phosphate rock
T.L. Roberts and M.D. Stauffer 70

Environmental issues in relation to cadmium in fertilizers
Tayeb Mrabet 79

Report of Session B 87

Session C

Accumulation in Agricultural Soils and Content in Food and Human Uptake

Issue Papers:

Current developments in the use of fertilizer phosphorus and the consequences concerning cadmium
G. Bertilsson 95

Cadmium in mineral fertilizers
Johannes Dettwiler 103

The role of farmers' organisations in reducing cadmium in food
Jan Eksvärd 104

Developing an Australian cadmium minimisation strategy
Graeme Evans 109

Sustainable cadmium management in agriculture: balancing the cadmium fluxes in arable land and grassland
Jens Folke and Lars Landner 113

Aspects of cadmium accumulation in agriculture
Simon W. Moolenaar, Panos Hatziotis and Theo M. Lexmond 119

**Cadmium accumulation in the soil –
an increasing problem**
Kierstin Petersson Grawé 133

**Fertilizer input of cadmium into Canadian
prairie soils**
T.L. Roberts 136

**The role of food regulations in minimising
exposure to cadmium**
Terry Spencer 140

Report of Session C 147

Session D

Uptake into Crops and Bioavailability

Issue Papers:

**Cadmium accumulation and availability in
agricultural land and the effects of land use changes**
Pierre del Castilho, Jan Bril, Paul Römkens and Oene Oenema 153

**Factors influencing cadmium content in crops –
results from Swedish field investigations**
Jan Eriksson, Ingrid Öborn, Gunilla Jansson and
Arne Andersson 168

**Management factors influencing cadmium
accumulation in crops**
C.A. Grant, L.D. Bailey and W.T. Buckley 169

Uptake of cadmium by crop plants
L.D. Bailey, C.A. Grant and R.J. Hill 174

**The relations between the cadmium content
in soil and in food plants**
Kimmo Louekari 180

Managing cadmium contamination of agricultural land
M.J. Mclaughlin, K.G. Tiller and A. Hamblin..189

Evidence for the leaching of surface deposited cadmium in agricultural soils
Fiona A. Nicholson, Kevin C. Jones and A.E. Johnston. 218

Report of Session D 237

List of Participants in the OECD Cadmium Workshop........................... 237

SESSION A

MEASURES AND TECHNIQUES TO REDUCE THE CADMIUM CONTENT OF FERTILIZERS

Managing the Cadmium Content of Phosphate Rock: A Contribution to Environmental Impact Mitigation

Arafat Ghoshesh, Saleh Bashir and Lana Dabbas

Jordan Phosphate Mines Company
Amman, Jordan

Summary

The Jordan Phosphate Mines Company (JPMC), conscious of the potential impacts on the food chain of cadmium released from phosphate rock and fertilizers derived therefrom, has devoted considerable efforts to mitigating the impacts by reducing cadmium content at the source. The techniques and measures applied for achieving this goal are discussed in this paper. They include selective mining, blending of rocks of different qualities, monitoring by extensive chemical analysis and elaborate geostatistical evaluations, carrying out appropriate sizing, and applying the results of intensive research.

JPMC activities aimed at cadmium minimisation go beyond the initial source materials. They also cover phosphoric acid and downstream products. JMPC's role in the World Phosphate Institute and its contributions to the Institute should also be recognised.

Introduction

There are no known separate ore deposits of cadmium. It is recovered mainly as a by-product of zinc refining and, to a small extent, of lead and copper refining. Cadmium is not essential for humans, animals and plants. It presents environmental risks. The human health risk is renal tubular dysfunction, which is manifest at certain critical cadmium levels in the renal cortex. When the body dose of cadmium is sufficiently high, a generalised disease results with anemia, bone pain, deformities and renal failure.

The cadmium problem became recognised during the last decade. The concern has become more important in recent times, mainly in some European countries but also in Australia and many Asian countries.

Cadmium is not biodegradable, and continuous accumulation in the soil is a distinct possibility if measures to reduce its presence to the minimum possible are not taken.

Cadmium is used in many products and processes, e.g. batteries, pigments, stabilizers, alloys, metal plating, neutron absorption, etc.

Soil can be polluted by cadmium from a variety of sources. These are mainly phosphate fertilizer, sewage sludge, solid waste, cement manufacturing, and other industries such as pigments, stabilizers, batteries, iron, zinc smelting, etc. Even according to the most stringent estimations and statistics, the share of phosphate fertilizer involved in soil contamination by cadmium does not exceed 7 per cent. However, great attention is given to this source in view of the dispersion over large geographical areas and the direct link to the soil-plant-animal and human cycle.

Certain authors estimate that about 80 per cent of the cadmium uptake by humans originates from the phosphate fertilizer source, whereas the contribution from this source to the total cadmium circulating in the environment is only 20 per cent.

Many regulations and measures have been issued in certain European countries and by international organisations, e.g. WHO. They include upper permissible levels for the cadmium content of fertilizer and sewage sludge and bans on certain uses of metallic cadmium. These measures have led to a reduction in annual additions of cadmium to soil in the EU from about 271 tonnes in 1985 to 251 tonnes in 1987, and below that last figure at present.

The cadmium content of phosphate fertilizer depends on that of the phosphate rock from which it is derived and the process used.

The cadmium content of phosphate rock of igneous origin is relatively low, generally below 5 ppm, but these rocks constitute only 15 per cent of world phosphate production. Sedimentary phosphate rocks contain between 3 and 120 ppm of cadmium.

At present, there is no feasible process for eliminating cadmium from phosphate rock or from the phosphoric acid used in fertilizer production. Any means, arrangement or manoeuver leading to a reduction in the cadmium content of phosphate fertilizer is therefore commendable. Activities carried out by JPMC are a contribution to the mitigation of the environmental impacts of cadmium.

Overview of the main processes of cadmium removal from phosphate rock and phosphoric acid

A. Cadmium removal from phosphate rock

As already mentioned, no really feasible process exists nowadays for the elimination of cadmium from phosphate rock in general. There could be an exception for rocks that have high organic matter content, in which case the organic matter can be reduced by calcination and the cadmium subsequently removed. Three main processes are presently encountered:

1. Cadmium removal by calcination using certain chlorine-containing additives, e.g. $CaCl_2$, $NaCl$, NH_4Cl, whereby cadmium is emitted in the form of $CdCl_2$. The efficiency of the removal reaches 100 per cent, in particular when $CaCl_2$ is used. The main disadvantage of the process is the necessity to eliminate the chlorine that remains in the phosphate concentrate because of a partial transformation to chlorine apatite. This results in an excessive corrosion of the equipment used for the phosphoric acid process.

2. Cadmium removal by calcination in an oxidizing atmosphere, whereby cadmium evolves in the form of CdO. The efficiency does not exceed 40 per cent at 1100°C.

3. Cadmium removal by calcination in a neutral or slightly reducing atmosphere, whereby cadmium evolves as metallic vapour, causing atmospheric pollution and hazard. This is the most significant procedure. It has reached the semi-industrial and even industrial scale (Nauru phosphates). The efficiency is approximately 75 per cent.

B. Cadmium removal from phosphoric acid

There are four types of processes for the removal of cadmium from phosphoric acid. With some of them phosphoric acid of high quality can be obtained. This phosphoric acid can be used in processes other than fertilizer production, e.g. feed and food stocks. None of these processes is applicable to the production of phosphoric acid for fertilizer.

1. Co-crystallisation of cadmium in $CaSO_4$ anhydrate

This sort of process is based on the fact that, owing to the similarity in atomic radius of cadmium and calcium, the first element can replace the second in different forms of hydrated calcium sulphate. The distribution of cadmium ions between calcium sulphate anhydrate and phosphoric acid is about a hundred times that observed in the case of the dihydrate. Therefore, cadmium can be removed more efficiently from phosphoric acid by co-crystallisation with the anhydrate. The latter is formed under conditions of high acid concentration in the first place and high temperature. There are two main processes.

The O.T.P./Becker process consists in adding a quantity of ground phosphate rock equivalent to about 2 per cent of the P_2O_5 content of the concentrated phosphoric acid and the stoechiometric quantity of sulphuric acid required to transform the rock into phosphoric acid. The mixture is heated to 110-120°C. Cadmium removal of 95 per cent is claimed for this process.

The CERPHOS process, developed at the Centre de Recherches des Phosphates at Casablanca, is based on the addition of sulphate ion in the form of gypsum to the diluted phosphoric acid before the concentration step. The removal of cadmium takes place at temperatures between 80 and 100°C and at a phosphoric acid concentration above 56.6 per cent P_2O_5. This process has many advantages over other processes and is retained for further development. The EU has funded additional research and development by about $US 1.5 million at the lab scale and 4.5 million at the pilot plant scale.

2. Cadmium precipitation by sulphide ions

Cadmium can be precipitated from phosphoric acid as sulphide by the addition of P_4S_{10}, Na_2S, NaHS, H_2S, etc. The best known application is the SIAPE process applied in Tunisia for the production of feed-grade phosphoric acid.

3. Cadmium removal by ion-exchange resins

Cationic ion-exchange resins, selective for cadmium, are used. Anionic resins, transforming cadmium into a complex anion with halogen, are also used.

4. Cadmium removal by solvent extraction

Use is made of several carbohydrate and amine solvents, such as octyl amine, Alamine-336, etc.

Cadmium mitigation in operations practiced at JPMC

A. Mining

JPMC operates three mines in the south of Jordan: El-Hassa, El-Abiad and Eshidiya. By the year 2000 Eshidiya, which is located 125 km north-east of the port of Aqaba, will be the major mining centre with a production estimated at nine million tonnes/year.

At Eshidiya, continuous phosphate-bearing areas and ore reserves in excess of two billion tonnes of phosphate were identified in detailed drilling and geostatistical studies. The phosphatic layers are:

A_1, to produce concentrate grade of 68-70 per cent BPL,

A_2, to produce concentrate grade of 73-75 per cent BPL, and

A_3, to produce concentrate grade of 75-77 per cent BPL.

Trucks transport the run of mine ores to screening and crushing plants, where the oversized low grade (over 12.5 mm) is removed. The ores below 12.5 mm are temporarily stockpiled in a linear storage according to the layers A_1, A_2 and A_3. Five stockpiles will be developed to separate A_1 ores into high and low CaO/P_2O_5 ratio, A_2 ore into high and low chlorine content, and A_3 ore.

The combined storage capacity of 65,000 tonnes will complement the selective mining operations to ensure the homogeneity of the ore for further processing.

Selective mining is practised in order to obtain the desired ore grades. It is usually carried out on the basis of BPL grade. Using the level of a trace element such as cadmium as a basis for selective mining is an innovation. It cannot be done without sacrificing certain rock grades that are otherwise good on the basis of their BPL grade. Geostatistical techniques allowing the establishment of averages for ore grades, bed thicknesses, total reserves, trace elements content, etc. are helpful in selective mining.

B. Preparation of phosphate ores

JPMC has invested millions of dollars in state-of-the-art blending facilities. A_2 screened ore will be conveyed to four 10,000-tonne storage silos to obtain a quality in accordance with customers' requirements. Four screened ore stockpiles, two for A_1 ore and two for A_3 ore, each of 26,000-tonne capacity, have been created to facilitate blending prior to beneficiation.

Quality specifications of the concentrates take into account the minimum acceptable P_2O_5 content and the maximum permissible levels of unwanted impurities.

A beneficiation plant with an output capacity of two million tonnes/year, comprising four washing lines, has been built. The flotation section comprises two lines for fine ores and one line for coarse ore.

C. Monitoring

Cadmium content is analysed in several laboratories belonging to the mines, the research department, the fertilizer complex, and the export department. An individual sample undergoes multiple analysis in the various laboratories. This allows close checking and verification of the results of the analysis. The samples are of different types: daily, weekly and monthly quality control samples; shipments; blending products; exploration; research; etc.

Atomic absorption employing the most recent and advanced spectrophotometers, including a new XRF (ARL 8480 S Sequential/Simultaneous Spectrometer), is used for the analysis of the samples. Data relevant to cadmium content are recorded and interpreted using the most recent computer techniques.

D. Sizing

Studies have shown that there is a good correlation between the cadmium and phosphate content of ores. This indicates that cadmium is mainly linked to the apatite crystal lattice, in which it replaces calcium ions. It has been noticed also that the cadmium content is relatively small in fractions lean in phosphate, which is the case of the coarse fractions. This, however, is not a general rule and many exceptions have been noticed. The reasons for these exceptions are the object of current research. It is hoped that, by finding out these reasons and by identifying and surveying the locations of beds that are exceptional in respect of distribution of cadmium in granulometric fractions, sizing could become a significant technique in cadmium reduction.

E. Phosphoric acid

JPMC produces phosphoric acid and diammonium phosphate fertilizer (DAP) at its fertilizer complex at Aqaba. The capacity for phosphoric acid is 1310 tonnes P_2O_5 /day, the major part of which is transformed to DAP. The cadmium content of the dilute phosphoric acid, the concentrated acid and DAP is respectively 3.5, 7 and 9 ppm. These values are well below the levels stipulated in the most stringent international regulations.

It is evident that at present, with the situation prevailing at JPMC, no cadmium reduction in phosphoric acid or DAP is needed. JPMC nevertheless takes part in all the activities of IMPHOS that are aimed at the removal of cadmium from phosphoric acid.

JPMC's internal research and possibilities of exploiting foreign research results

Numerous research activities aimed at cadmium reduction are going on or planned at JPMC in order to keep the company up-to-date with developments and changes that might take place in the world regarding cadmium legislation. Some examples are given hereafter:

- elaborate mineralogical studies concerning the status of the cadmuim ion in the apatite lattice, distribution in different forms of apatite, major impurities and organic matter, relation to CO_2 and content of arsenic, etc.

- surveys of cadmium distribution in granulometric fractions of ore in different locations to locate places where the coarse fractions are enriched in cadmium and to identify the reasons.

- identification of factors that enhance the displacement of cadmium into phosphogypsum, and application of those factors in a way that does not compromise the quality of the produced phosphoric acid.

- several studies, in particular on Eshidiya phosphate, were carried out to establish correlations between cadmium, arsenic and other elements.

The Eshidiya mines are characterised by the existence of two distinct areas, coquina and non-coquina. In the coquina area a thick layer of coquina limestone lies on top of the phosphate layers. In the non-coquina area the coquina limestone has disappeared due to a lateral facies change of limestone to marl. There are major differences in chemical characteristics between the ore of the two areas.

The ores of the coquina area have the following basic characteristics:

- moderate to high calcium oxide content in A_1

- low content of siliceous materials in the A_1 and A_2 layers

- low content of chlorine, aluminum and iron oxides

- low content of cadmium and arsenic.

The ores of the non-coquina area have following basic characteristics:

- low CaO/P_2O_5 ratios in the bearing phosphate particles
- siliceous material present as abundant red and yellow clays and free quartz
- relatively high content of chlorine, aluminum and iron oxides
- relatively moderate content of cadmium and arsenic.

JPMC has decided to investigate in depth the correlation between cadmium and other elements in rock produced from the Eshidiya A_2 layer. A joint research programme has been established with Hydro Supra Company to this effect. Through statistical evaluation, the following correlations were established:

- The lowest cadmium content is found together with a low Al_2O_3 content.
- The lowest cadmium content is found together with a low Fe_2O_3 content.
- A low cadmium content is found together with a high carbonate content.

The following weak correlations for both cadmium and arsenic were found:

- The lowest cadmium and arsenic contents were found together with low Al_2O_3 and Fe_2O_3 contents.
- Low cadmium and arsenic contents were found together with high carbonate contents.
- The lowest cadmium and arsenic contents were found together with low silica content.
- Low cadmium and arsenic contents were found together with high P_2O_5 content.
- Low cadmium and arsenic contents were found together with high CaO content

JPMC's research staff is following all results of foreign research related to cadmium risk reduction. They are always trying to test, evaluate and comprehend the merits and implications of this research.

References

1. A. Davister (1992) "Inventaire des études et recherches sur les procédés d'élimination du cadmium dans l'acide phosphorique."

2. Proceedings of the Workshop "Cadmium Removal from Phosphoric Acid," Casablanca, 13-14 July 1993. Imphos.

3. G. Baudet (1992) Etude documentaire sur les possibilités d'élimination du cadmium à partir de concentrés de phosphate. BRGM, R 35890.

4. P.A. Maxson and G.H. Bonkeman (1992) Les Métaux Lourds dans les Phosphates. Risques pour l'Environnement et Implications Stratégiques. Commission des Communautés Européennes, Bruxelles.

5. S. Bashir, et al. (1993) Cadmium removal from phosphoric acid and from phosphates by solvent extractions and by other direct methods or via acid purification (in Arabic). Proceedings of Jordanian Chemical Engineering Conference, I, October 18-20, 1993, Amman.

6. Draft OECD Report (1991) Co-operation on existing chemicals: risk reduction: lead country report on cadmium. Published in 1994 as Risk Reduction Monograph No. 5: Cadmium.

7. Rapport final CERPHOS (1995) Travaux Complémentaires de Recherche sur le Procédé CERPHOS. Projet de décadmiation de l'Acide Phosphorique.

8. A.J. Williams (1992) The development of environmental legislation in Europe and its impact on the market of phosphates fertilizers. In "Proceedings of and International Workshop on Phosphate Fertilizers and the Environment," March 23-27, 1992, Tampa, Florida, U.S.A. IDF Publications.

9. Arafat Ghosheh and Faisal Dodeen (1993) Heavy Metals in Jordan Phosphate Rock, Fourth international fertilizer seminar, June 12-18, 1993, Amman.

10. Anders Axelsson and Goran Jonsson, Statistical evaluation of analytical data from A2 layer of Eshidiya mine in Jordan, Hydro Supra.

Studies and Research on Processes for the Elimination of Cadmium from Phosphoric Acid

A. Davister

Liège, Belgium

Foreword

At the request of IMPHOS (Institut Mondial du Phosphate), the CEC (Commission of the European Communities) hired me to update the information available on this subject as of 1992. Later on, I made another update with the view of writing a paper for the IFA (International Fertilizer Industry Association) technical meeting in Amman, Jordan (October 1994). The CEC asked me to supervise a process development made by IMPHOS in 1994-1995 with CEC funding.

These are the activities which gave me the background to write this paper.

Aspects of the problem

The supply of phosphorus to agricultural soils is essential to compensate the quantities exported by crops, but it happens that phosphate ores contain variable quantities of cadmium (Table 1). As you can see, these cadmium contents are quite low, although they cover a wide spectrum. An estimation made in 1989 by EFMA (European Fertilizer Manufacturers Association) put the average cadmium content in European fertilizers at 60 mg per kg P_2O_5, which is 500 times less than the soil cadmium content at the fertilizer's usage rate in 1989 (Table 2).

But fertilizers are not the only cadmium source for soil. The fall-out and by-products of human activities bring more cadmium to the soil than do fertilizers (Table 3).

The input of cadmium to the soil may exceed export by plants. As a consequence, the cadmium content of the soil will increase. This does not necessarily mean that more cadmium will proceed into the food chain, as is discussed in other parts of this Workshop.

Nevertheless, some people are of the opinion that the increase in soil cadmium content should be stopped immediately. They generally want to fight all cadmium sources equally.

In my opinion this is not fair. While fall-out and by-products of human activities should be controlled by all possible means, the fact that phosphorus is essential in feeding mankind justifies its cadmium input being controlled only within economically sustainable limits (any grabbing spirit apart).

Table 1

Average P_2O_5 and cadmium contents of the main commercial phosphate rocks

Phosphate rocks	P_2O_5 % Total	Cadmium		
		PPM/total	PPM/P_2O_5 or mg/kg P_2O_5	PPM/P or mg/kg P
1. Igneous origin				
Kola	39	<5	<13	<30
Pharlaborwa	37	<5	<13	<30
2. Sedimentary origin				
Florida	32	7.5	23	54
Jordan	33	<10	<30	<70
Khouribga	32/33	15	46	106
Syria	31	16	52	119
Algeria	29	17.5	60	138
Egypt	27	20	74	170
Bu-Cra	34	34	100	229
Nahal Zin	31	31	100	229
Youssoufia	33	40	121	277
Gafsa	29	40	137	315
Togo	37	60	162	371
North Carolina	30	50	166	381
Taiba	37	75	203	464
Nauru	37	90	243	530

Table 2

Cadmium in soil from fertilizers
(EFMA figures, 1989)

1) **Average cadmium content in European soil**

    or 0.5 mg / 500 µg per kg of soil

2) **Average cadmium content in European fertilizers**

    60 mg per kg of P_2O_5

3) **Average amount of cadmium in European soils from fertilizers**

    2.5 g of cadmium per hectare per year

    or

    1.0 g of cadmium per acre per year

4) **Time required to double cadmium content**

    (if no cadmium was exported from the soil)

    500 years

Table 3

Sources of cadmium brought to the soil in Europe
(EFMA figures, 1989)

1)	Atmospheric (rain and dust)	
2)	Sewage sludges	
3)	Animal sludges	350-400 tonnes cadmium
4)	Atmospheric (rain and dust)	250 tonnes cadmium

Rich and relatively sparsely populated countries should not impose strict regulations that fit their conditions but will only mean more economic constraints for poor and/or densely populated countries in the developing world. In fact, if the limitation of fertilizer's cadmium input had to be obtained by ore selection, as it is the case now, this would exclude many ores (Table 4). There is no doubt that this would make the phosphorus resource insufficient to satisfy the needs of mankind. Moreover, it would deprive some developing countries of much needed cash flow.

Phosphate industry reactions

This situation does not mean that the phosphate mining and processing industry does not care about the cadmium conundrum; in fact, there has been much interest in cadmium elimination processes, but up to now none of these processes has reached the industrial stage under conditions which are economically sustainable for the fertilizer industry.

Generally speaking, one can summarise the world situation as follows:

- The North American industry has no problem, as cadmium in fertilizers is not an issue on that continent.

- The European industry is not in a position to solve the problem, although it is confronted with regulations varying from country to country (Table 5) and it tries to cope with the regulations by selective phosphate supply.

- The African industry, which is engaged in full development of its processing activity, is very concerned with the subject and has lately shown a lot of interest in research and development of cadmium elimination processes.

While cadmium removal from phosphate rock has not up to now given way to any promising research route, the decadmination of wet process phosphoric acid has been researched in many ways. They can be grouped as shown in Table 6.

Table 4

Permitted rate and phosphate exclusion (mg Cd/kg P_2O_5)

Permitted rate	Phosphates excluded
250	none
200	Nauru, Senegal
150	Nauru, Senegal, Togo, North Carolina
100	Nauru, Senegal, Togo, North Carolina, Tunisia, Morocco (Youssoufia), Israel (partially)
50	Nauru, Senegal, Togo, North Carolina, Tunisia, Morocco (Youssoufia), Morocco (Bu-Cra), Israel (totally), Algeria, Egypt, Syria

Table 5

Maximum concentrations of cadmium allowed in phosphate fertilizers (mg Cd/kg P_2O_5) (from the OECD Risk Reduction Monograph on cadmium, May 1994)

Country	Limit
Austria	120
Belgium	90
Denmark	110
Denmark (from July 1995)	65
Finland	22
Germany	90 (voluntary)
Japan	150
Norway	44
Sweden	44
Switzerland	22

Table 6

Grouping of decadmiation processes for wet process phosphoric acid

	Principle of the process	Identification of the route
1.	Co-crystallisation of cadmium with anhydrite	**CC**
2.	Precipitation of cadmium by molecules containing sulphide ion	**PP**
3.	Removal of cadmium by an ion exchange	**RX**
4.	Removal of cadmium by solvent extraction	**SX**

The most recent development

In view of the situation after the 1992 survey, my conclusions were as follows:

- If cadmium in fertilizers happened to be a problem, it was a long-term one.

- As there was ample time available, I suggested that no general regulation be implemented before more in-depth studies were performed with the main two objectives of:

    a) developing knowledge about bioavailability of soil cadmium through the food chain;

    b) developing the most promising route(s) of cadmium elimination from phosphoric acid.

The first objective has been dealt with in other papers of this Workshop. Here I will treat the second objective. In view of an interesting project submitted by IMPHOS, the CEC has allowed it a credit of 1.1 million ECU to proceed to the first phase of development of its CC route.

After installation of the appropriate equipment, the lab-scale pilot plant (capacity: 25 litres of acid per hour) began operation in mid 1994 at the CERPHOS premises.

The tests are now completed and have been successful in determining a modular decadmiation process able to solve all cases from the simplest to the most complicated, depending on the final use of the acid and on local environmental constraints.

This process consists of three main steps:

- Cadmium elimination:

  This is the most important step. Cadmium present in the feed acid is eliminated by cocristallisation into a precipitate of anhydrite of calcium sulphate, and separation of the latter occurs by settling and/or filtration. Starting with a feed acid containing 75 mg cadmium per kg of P_2O_5; an acid has been consistently been produced with less than 10 mg cadmium per kg of P_2O_5.

- Product acid desulphatation:

  The precipitation of calcium sulphate anhydrite requires the addition of sulphuric acid to the feed acid. If the cadmium-free acid is used to manufacture regular fertilizers like TSP or DAP, there is no need to remove the additional sulphuric acid. But its removal is required for some acid uses, and the conditions of this operation have been determined.

- By-product post-treatment:

  Removing the cadmium from the acid is carried out by precipitating it into the anhydrite in a more concentrated form than in the original ore. No further post-treatment is required if one has a safe disposal area. But, in some locations, the by-product has to be concentrated to be disposed of as a hazardous or toxic waste. Research has shown that it is possible to concentrate the cadmium-bearing by-product into the most concentrated and least toxic residue: cadmium metal itself.

These are the "chemical" developments achieved. But the first phase included another goal – an "engineering" goal that has been completed, too.

Thanks to the information collected during the test, it was required to design the second phase, i.e. the semi-industrial pilot plant. These engineering works have incorporated:

- the selection of pilot plan capacity, which has been set at 3 m^3 per hour of 30 per cent P_2O_5;

- the flow sheet of the pilot plant;

- the mass and thermal balances of same;

- the list of equipment;

- the preparation of the files required to be included with the invitation to bid;

- the estimation of construction and operation costs and timing.

This last item is the most able to attract the interest of this Workshop.

Let us summarise briefly by saying that the pilot plant costs are estimated at:

- $US 6 millions for the investment;
- $US 3 to 4 million for the operation.

The timing is estimated at:

- two years for construction (including bidding time);
- around two years for the operations, in view of optimising all the phases of the process and checking them with two or three other acids.

Costs estimation

To the best of my existing information, the investment and operation costs relative to the different process routes can be estimated as shown in Table 7. For the convenience of the reader, we have included in this Table the comparative costs for the wet process phosphoric acid plant itself.

As a consequence, the appraisal of those routes can be made as follows:

CC - seems the most promising route.

One process has reached the lab-scale pilot plant stage, and the engineering file for the semi-industrial pilot plant is ready.

All costs are estimates by professional engineering personnel, and this shows that investment costs are +/- 30 per cent of the wet acid costs while the operation costs may vary between 20 and 30 per cent of same.

PP - This route is characterised by excessively high operating costs due to the use of expensive reagents.

There is one industrial plant in operation in Tunisia (capacity: 100,000 tonnes of P_2O_5 per year), but it manufactures feed grade products which have a higher added value than the fertilizers and so are able to bear a higher acid cost.

The investment costs, which are estimates by development people, might be kicked up when estimated by engineering professionals. The present value represents 30 per cent of wet acid, while the operation cost is 75 to 100 per cent of same.

RX and **SX** both exhibit excessively high investment and operation costs due to expensive pre-treatment of the feed acid and use of high-value solvents or resins. There is one industrial plant in operation in Germany using the SX route (capacity: 60,000 tonnes of P_2O_5 per year), but it operates in association with high-value food grade acid production, which bears the full cost of feed acid pre-treatment and can afford more expensive acid price.

Table 7

Summary of investment and operation costs (in millions of $US) of various processes for decadmiation of phosphoric acid in a 500 tonnes/day P_2O_5 unit located in Western Europe

Type	Processes Conditions	Investment	Cost (per tonne P_2O_5)
CC	Without residue treatment	7	6
	With residue treatment	8	6-9
PP		7	30
RX	With medium acid pre-treatment	9	30
SX	With full acid pre-treatment	9	32
Wet process phosphoric acid manufacture		25	30-40

Important remarks:

1) At this stage of process development, these costs have only a relative and indicative value since they may increase substantially as a result, for example, of substantial P_2O_5 losses or difficulties in the residue treatments.

2) To value these costs, you should compare them with the costs associated with wet process phosphoric acid manufacture:

- *The investment costs range between 30 and 40 per cent of acid manufacture (the percentage may still increase as research progresses);*

- *The operation costs range between 20 and 100 per cent of acid manufacture.*

The same remarks as for PP above apply to the investment costs. The present values represent roughly 40 per cent of the wet acid plant, while the operation costs are 75 to 100 per cent of same.

Conclusion

For the regular fertilizer grade phosphoric acid, the CC route seems by far the most promising, with estimated investment costs in the range of 30 per cent of the wet acid plant of the same capacity and operation costs ranging between 20 and 30 per cent of same. Nevertheless, this route has not yet been developed further than the lab-scale pilot plant stage. To determine if it is able to deliver its promises at a sustainable cost, it is necessary to operate it at the semi-industrial pilot plant stage. This development is prepared, and is estimated to cost $US 10 million and last four years.

If the important financial and human resources needed are available for this project, the answer concerning its real possibilities might be available by the end of the century.

Cadmium Removal from Phosacid

R.M. Vermeul

Hydro Agri Rotterdam
Vlaardingen, The Netherlands

Introduction

In 1988, Hydro Agri Rotterdam (HAR) decided to invest in conversion to two filter hemi-dihydrate (HDH) processing for one of its phosacid plants.

HAR operated two single-stage hemihydrate (HH) plants based on Hydro Fertilizer Technology know-how. Overall P_2O_5 efficiencies were in the normal range for single-stage processing and varied from 92 to 94 per cent.

HAR is located in the Rotterdam port area, and the gypsum produced is discharged as a slurry into the Rhine river about 20 km from the sea.

In 1988, the Dutch government and HAR agreed a covenant. In this covenant HAR committed itself to a phased decrease in emissions of phosphate, cadmium and other heavy metals.

For cadmium, a fixed end level of max. 0.6 tonne/year and 0.5 mg/kg gypsum was agreed, though the technology to achieve this was not yet available. For P_2O_5, a fixed reduction of at least 50 per cent had to be reached. For the other impurities, high reductions were set as targets.

The decrease in emissions with the gypsum should not lead to an increased environmental burdening by end products based on the intermediate phosphoric acid.

For 1994, a maximum level for fertilizers of 55 mg Cd/kg P_2O_5 was fixed and a value of 15 mg cadmium/kg P_2O_5 for fertilizers was included in the covenant as a goal for the year 2000.

HAR's feedphosphate production requires a level of lower than 20 mg cadmium/kg P_2O_5.

To remain flexible with regard to rock choice, purification of phosphoric acid - especially from cadmium - had to be investigated in addition to the purification of the gypsum.

To achieve both ultimate objectives, HAR conducted a comprehensive research programme. For cadmium removal from phosphoric acid, three techniques were investigated on a pilot plant or bench scale.

Liquid-liquid extraction and ion exchange were investigated by HAR. Co-precipitation was studied in co-operation with P. Becker, France, and Office Togolais des Phosphates (OTP); results on co-precipitation were also presented at the 1992 IFA conference.

In 1994, the three techniques were evaluated during the IFA conference in Amman.

This paper is an abstract from the Amman presentation.

In December 1991, start-up of the converted HDH-plant took place.

Presently Jordan rock is used, as it is sufficiently low in cadmium to produce phosphoric acid directly complying with feedphosphate standards.

Towards a clean gypsum: the two-filter HDH-extra process

Description

HAR was aiming for a very low cadmium emission, on which criterion all existing technologies fell short.

To achieve acceptable cadmium emissions, the standard HDH-process had to be improved. In Figure 1, a flow-chart of the HDH-extra process is shown.

The main process characteristics of the HDH-extra process are a low emission of cadmium, phosphate and heavy metals by a high degree of conversion of hemi- into dihydrate gypsum and chloride-addition.

Emission of cadmium for different phosphoric acid processes

In Figure 2, cadmium concentration in the gypsum is plotted versus the cadmium-concentration in the rock for different process types.

The dotted line depicts the critical level in the gypsum, which had to be achieved: 0.5 mg cadmium/kg gypsum.

For HAR applicable rock types vary in cadmium concentration from 6 to 50 mg/kg Cd. It is clear that even with a rock with 6 mg Cd/kg rock, both the HH and the standard HDH process will render a too high cadmium emission.

Figure 1

The HDH-extra process

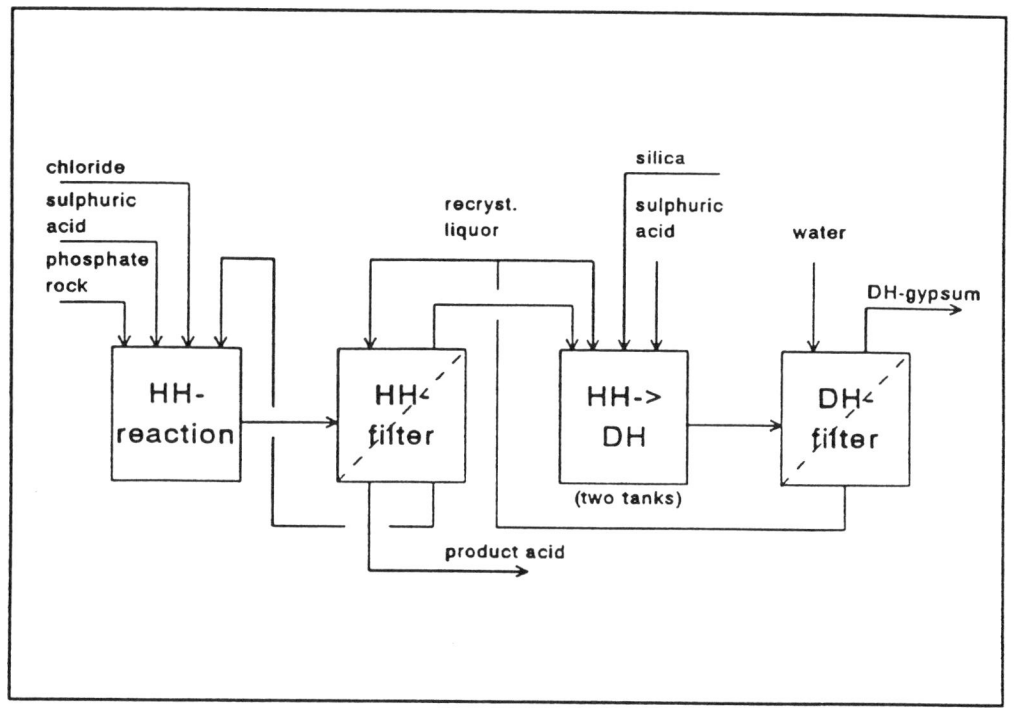

Figure 2

Effect of cadmium concentration in rock on the cadmium concentration in gypsym, for different processes

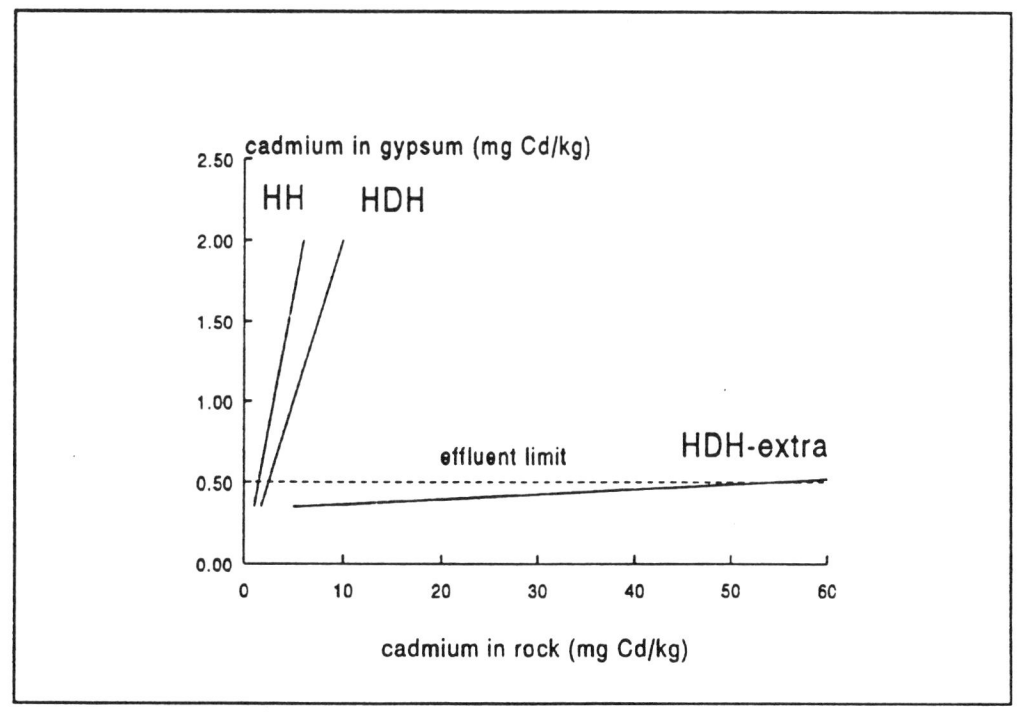

Because of this, HAR had to develop the HDH-extra process, of which the cadmium emissions are also shown in Figure 2.

Key to the drastically decreased cadmium emission is the addition of chloride to the HH-stage. Chloride complexes with cadmium and these complexes are not incorporated in the gypsum. The HH-gypsum, and consequently the recrystallized DH, thus contain less cadmium.

Purification of HDH-extra phosphoric acid: cadmium removal

Introduction

As the Norsk Hydro HDH-extra process was primarily developed to produce a clean disposable gypsum, an increase of impurities in the only other process-outlet - the phosacid - will occur.

After a first selection and evaluation two decadmiation techniques were selected and submitted to pilot plant tests while a third test unit was operated in co-operation with OTP/Becker.

Simultaneously it was studied how cadmium could be obtained in a concentrated form ready for deposit or recycle.

When studying the "state of the art" regarding cadmium removal, it became clear that there was at the time no full-scale process running on crude product acid, called black acid.

Cadmium-removal techniques: a first selection

In this section the cadmium removal techniques as reported in literature are described, together with the applicability in HAR's specific situation.

The highest cadmium concentration to be treated was defined at 65 mg Cd/kg acid.

Extraction

Solvent extraction is based on the principle of dissolving a cadmium scavenger in an organic solvent, immiscible with phosacid. When stirring the organic/acid mixture, cadmium is transferred into the organic phase. As soon as stirring stops, both phases separate.

The organic phase is subsequently decadmiated by washing with an aqueous phase, decomposing the scavenger cadmium complex. The regenerated organic phase is reused, while the aqueous cadmium concentrate can be processed further.

Hydro Supra in Sweden tried in a pilot plant to purify black phosacid according to this principle and discovered that crud formation (see below) could seriously disturb operations as it hinders separation of the organic and aquous phase.

Hence Supra developed acid pre-treatment to limit this problem, but after some time stopped the development.

The organic compounds used extract anions. The presence of chloride in 42 per cent P_2O_5 HDH-extra acid will automatically supply negative cadmium chloride complexes.

Besides, being a negative factor (corrosion), the high Cl concentrations favour the extraction technique

Ion exchange

Ion exchange is based on the use of a cadmium scavenger fixed on the surface of small porous beads (a resin). These beads are loaded and subsequently regenerated, similar to the two basic steps of extraction. Two types of resin were evaluated.

Cation exchange (CE)

CE was soon abandoned because the resins adsorb most bivalent ions (Ca, Fe, Mg, etc.) preferential to cadmium. As these other ions are present in excess compared with cadmium, selectivity will be poor, resulting in large columns and a large regenerate flow to be treated.

Anion exchange (AE)

AE looked promising since cadmium is dominantly present in HDH-extra acid as a negative cadmium chloride complex ($CdCl_3$), and in addition shows higher affinity for the resin than competing anions like sulphate and chloride.

Anion exchange (AE) was hence selected for further investigation.

Inorganic sulphide precipitation

From the literature it is known that inorganic sulphide salts have very low solubility products. In phosphoric acid, however, the actual S^{2-} concentration that can be reached - needed to form cadmium sulphide - is very limited due to the high acidity which promotes H_2S formation.

As a consequence, high pressure is required to increase the S^{2-} concentration. Another method is saturating a carrier with a sulphide solution and subsequently contacting this carrier with acid.

As expected, the "natural" presence of cadmium chloride complexes in HDH acid proved to be a disadvantage for this technique.

Though these methods worked in the end, they proved to be difficult to control and were abandoned.

Organic sulphide precipitation

Some commercially available organic sulphides were tested on concentrated acid with regard to their potential cadmium removal efficiency. These compounds usually contain a thiophosphate group, whose properties can be compared with S^{2-}.

However, when confronted with the smell of these compounds and their known toxicity, development was stopped.

Still, some acid producers use this method to produce a low-cadmium acid.

Electrolysis

A major advantage of electrolysis could be the direct separation of cadmium as a metal, which even has some commercial value.

In practice, realization is not yet feasible:

- Suppliers of these cells do not give guarantees: the membrane cells required still have to be developed.

- Energy efficiency in 42 per cent P_2O_5 acid is extremely low due to the high acidity: competition between H^+ and Cd^{++} will cause strong hydrogen-formation.

- The presence of chloride in the acid results in low concentrations of free Cd^{++}.

An estimated investment of approx. 20 MLN USD for a unit to treat 100,000 t P_2O_5/year made this technique unsuitable for practical application within reasonable term.

Co-precipitation with calcium sulphate anhydrite ($CaSO_4.0aq$)

The cadmium uptake with different calcium sulphate modifications can be summarized as follows: relative uptake in DH : HH : AH = 1 : 4 : 200.

The strong uptake in anhydrite can in principle be used to remove cadmium from acid by adding a calcium source to concentrated phosacid at elevated temperatures.

During HAR's experiments, it was found that the uptake was relatively low, implying production of large volumes of anhydrite which cannot be disposed of.

The anhydrite must consequently be reconverted into dihydrate to liberate the incorporated cadmium and obtain a clean DH gypsum.

Conversion into dihydrate, however, would require days unless unpractical conditions (temperature and acidity) were applied. This route did not seem technically realistic for HAR.

See also below.

Pilot plant research

Eventually two techniques, extraction and ion exchange, were selected and investigated at HAR in different pilot plants because the need for acid purification was expected within a few years. Separately OTP/Becker approached HAR to test their version of the co-precipitation process on our site. Their results are mentioned briefly, as they were reported earlier.

HAR's pilot plant results are described in the following paragraphs.

Extraction

HAR could make a quick start with research on extraction, as a pilot plant was still present at the site of her sister company, Hydro Supra, Sweden.

The 100 l/h pilot plant was transported to Vlaardingen. A computer system enabled automatic start-up, process-control and automatic stop in case certain process values were exceeded.

High concentration factors (25-50 m^3 acid/m^3 regenerate) and low cadmium levels in the treated phosacid (< 1 mg/kg) were obtained because the cadmium uptake in the organic phase increases strongly with the Cl concentration in the phosacid.

Crud formation was mentioned above. Crud is a third viscous phase that can be formed during extraction. It is known to disturb the extraction process by accumulating at the interface between acid and organic phase.

This effect hinders phase separation and enhances entrainment of organics with the end product.

Organic compounds in black phosacid (humic acids) are responsible for crud formation. Removing these compounds with adsorbents at 50-60°C reduces crud formation considerably.

Another phenomenon - common in extraction processes - is entrainment. After settling, the acid will contain about 100 mg/kg of solvent if no post-treatment is installed.

Furthermore, extraction requires explosion-proof equipment and airtight extraction compartments to prevent evaporation of organic solvent.

Techniques like co-precipitation and ion exchange evidently do not show these negative effects nor do they require extensive pre- and post-treatment.

Anion exchange

A special pilot plant to assess ion exchange (IE) was obtained from the American company "Advanced Separation-Technology" (AST): a carousel type (60-100 l/h). Instead of stationary columns, this type of equipment uses a rotating carousel in which many small columns are placed (Figure 3).

It is obvious that, with IE, entrainment and crud formation will not occur, so post-treatment and solvent removal from phosacid will not be required.

The cadmium uptake proved to be sufficient to establish low cadmium levels in the acid at a reasonable resin loading.

It was surprising to find that the columns were less sensitive towards blockage than usually assumed; acid pre-treatment by settling proved to be sufficient, but cannot be omitted.

By upflow loading and downwards rinsing of the columns, solids originally present in the acid are washed down. Regeneration with water proved to be effective.

After 1500 hours of operation, no attrition of the resin or damage by osmotic or thermic shocks could be deduced from parameters like pressure drop and resin capacity or from visual (microscopic) inspection.

During operation of the pilot plant, no unexpected drawbacks to this process were encountered. The cadmium removal process, including equipment, could be kept well under control. The presence of chloride never caused problems, as proper construction materials were selected.

Co-precipitation

Apart from ion exchange and extraction pilot plants, a bench-scale unit was operated on HAR's site by Becker/OTP. The results of their investigations were reported during the IFA conference in The Hague (1992). It is worth mentioning that both cadmium uptake and the recrystallization to DH gypsum were remarkably optimized during this study.

Cadmium disposal

All three methods produce a liquid regenerate which needs further treatment to obtain a cadmium concentrate.

The simplest and most effective way to remove cadmium from the liquor is precipitation with lime or sodium hydroxide, followed by filtration. The filter cake has to be deposited.

Three other methods have been developed to produce rather pure cadmium residues that can be recycled to zinc-producing industries. The value of these products does not, however, compensate for the required investment and operational costs.

Figure 3

Carousel unit for ion exchange

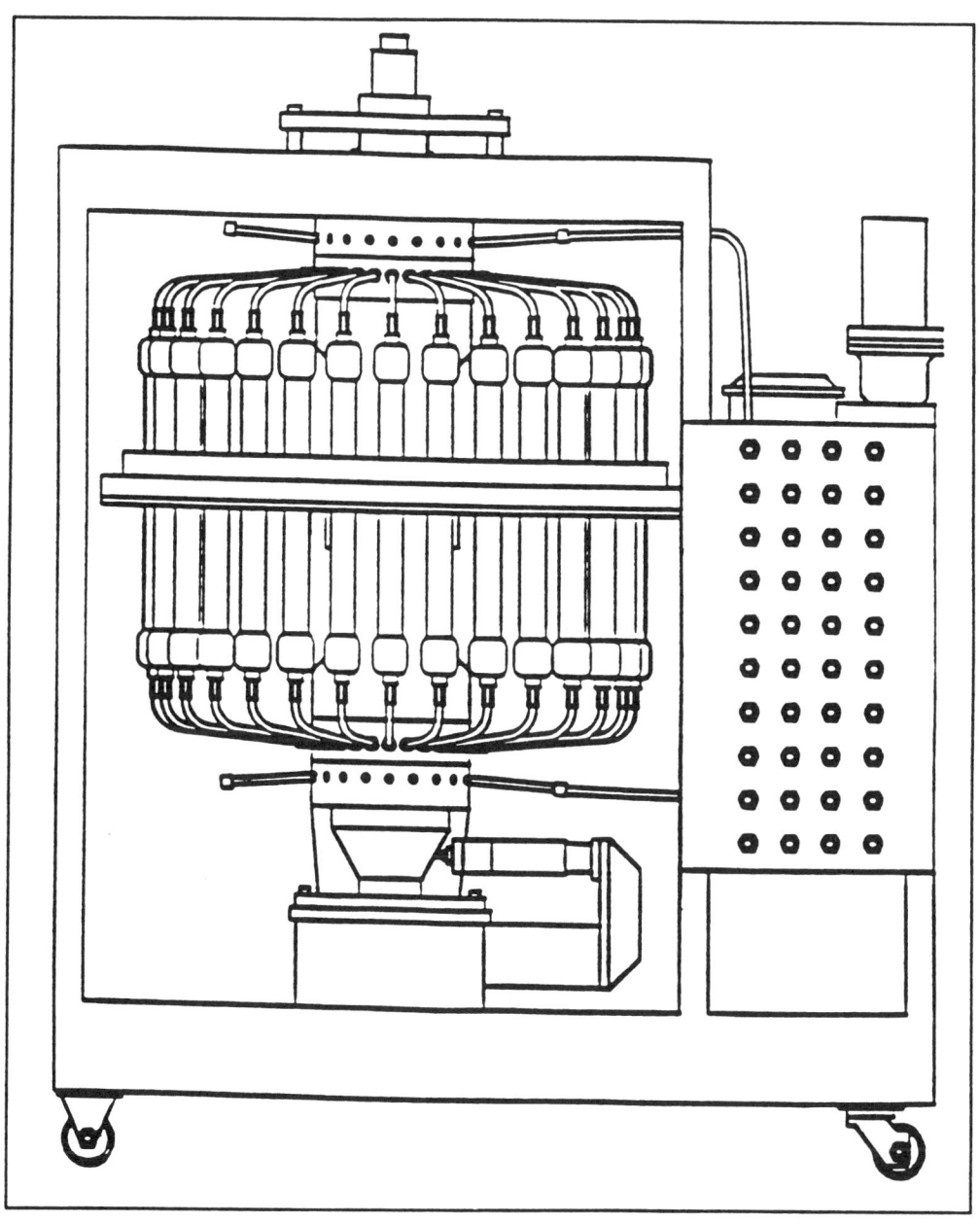

For HAR, precipitation is therefore considered to be the favoured case.

Evaluation

The HDH-extra process, a result of intensive R&D, produces phosphoric acid in an environment-friendly way. Emissions of phosphate, cadmium and other heavy metals are drastically reduced.

With this reduction the concentrations in the phosphoric acid increased, and therefore cadmium removal has been investigated, operating two different decadmiation pilot plants and co-operating with bench-scale research by Becker/OTP. The goal of 12 mg Cd/kg P_2O_5 (and even lower concentrations) can be met by each technique.

For HAR, ion exchange is the most appropriate technique to remove cadmium from phosphoric acid. This technique is simple, reliable, and does not require extra process steps like pre-treatment, post-treatment, crud removal, etc.; pre-settling will probably suffice.

The fact that, in the HDH-extra process extra chloride addition is required for high-cadmium rocks fits perfectly with the same need during ion exchange in order to obtain a low cadmium acid while producing a small volume of regenerate.

The resin can simply be regenerated with water. The regenerate is finally decadmiated by precipitation followed by filtration, and the P_2O_5-containing filtrate is recycled to the HDH plant.

Total cost of cadium removal depends on the rock origin and amounts to between $US 15 and 25/tonne P_2O_5.

The investment for a 125,000 tonne P_2O_5 unit is estimated at $US 3.5 million.

Promoting the Development and Semi-Industrial Application of a Potentially High Performing Process for Cadmium Removal from Phosphate Rock

Abdelaâli Kossir and Abdellah Chik

**Groupe Office Chériffien des Phosphates
Centre d'Etudes et de Recherches des Phosphates Minéraux
(CERPHOS)**

Considering the growing concern for environmental protection, the phosphate producing countries and some of the sponsors of industrial processes have sought methods for cadmium elimination from phosphate rock.

However, there is as yet no economically viable industrial process for reducing the trace amount of cadmium in phosphate rock or phosphate rock concentrates to an acceptable level.

The Center for Research on Phosphate Ores (CERPHOS) in Casablanca, Morocco, was directed by the OCP Group, as part of its environmental policy, to conduct research on cadmium elimination from phosphate rock. CERPHOS developed a process for cadmium removal by thermal treatment of the rock and trapping of volatilized cadmium in solid residues in a stable form.

This paper is intended to present, and show the interest of, the process.

Brief description of the CERPHOS process for cadmium removal from phosphate rock

Thermal treatment of the rock

The CERPHOS process permits the removal of cadmium and sulphides from phosphate rock and rock concentrates, using a thermal treatment which proceeds in two steps:

- First step: reduction and volatilization of cadmium under controlled pressure and a temperature ranging between 750°C and 950°C;

- Second step: oxidation of sulphides generated during the first step, in addition to oxidation of the residual organic matter, both at a temperature ranging between 700°C and 850°C.

The procedure was selected as a function of the type of rock or rock concentrate to be treated (content of cadmium, organic carbon and carbonates). The relatively low temperatures used minimize the risk of aggregate formation and conserve to a large extent the inherent acidulation capacity of the rock.

Phosphate rock treated in this way contains less than 15 per cent of its initial cadmium concentration and has only traces of sulphides and organic carbon.

Treatment of cadmium-containing off-gases

The hot cadmium-containing gases resulting from the first step (thermal treatment) are oxidized and cooled by injection of an optimal air dose. Cadmium is then precipitated in cadmium oxide form and left to settle on smoke particles or particles injected for the purpose. Dust is removed from the gases, which are recycled in the process at a temperature sufficiently high (T $>300°C$) to permit the recovery of heat and the improvement of the energy cost-effectiveness of this thermal treatment.

The cadmium ends up being trapped in a stable form in a solid residue which can then be processed and stored at minimum risk to the environment.

Conclusion and future research needs

The CERPHOS process for cadmium removal from phosphate rock offers the mining industry an effective method for decreasing the level of cadmium in its phosphate rock:

- The process permits the production of low-cadmium rock phosphate from high-cadmium feed phosphate. The treated rock is also low in sulphides and organic matter.

- The process functions by trapping the volatilized cadmium in solid residues under a stable form.

- The gas heat is recycled to a large degree in the process, improving as such the energy cost-effectiveness of the thermal treatment.

In order to demonstrate the technical and economic feasibility of this potentially high- performing process, it is indispensable to take the next step of testing the process on a semi-industrial pilot scale.

There are several justifications for the pilot tests:

- collection of technical and economic data required for a feasibility study, and design of an industrial unit for cadmium removal from phosphate rock;

- optimization of circuits, equipment and materials in a way that reduces to a minimum the risks of aggregate formation;

- determination of the applicability of the process to other phosphate rocks.

The construction and implementation of a pilot plant operating on the principles of this process, and the expenses of a pilot experimental program, were both estimated to cost $US 15 million.

The implementation of the pilot plant and the conduct of experimental work would require a time period of five years.

The cost of cadmium removal from phosphate rock is estimated at about $US 10 per tonne of phosphate concentrate, based on a plant unit capacity of 400 tonnes of rock per day.

REPORT OF SESSION A

MEASURES AND TECHNIQUES TO REDUCE THE CADMIUM CONTENT OF FERTILIZERS

The papers presented in this section provided the basis for the discussion.

The present demand for low cadmium phosphates and fertilizers

Demand for phosphatic fertilizers is increasing throughout the world. This applies equally to both lower and higher cadmium content product. At this point there has been no great shift in the overall market share of the two categories. However the change to lower cadmium content products has been very significant in some countries and this trend might take on greater importance globally in the future.

Factors affecting demand

1. Access to low cadmium rock sources and long-term trade agreements with phosphate-producing countries.

2. Environmental concerns vary from country to country, as do soil and climatic conditions, types and varieties of crops used, and farming practices, all of which might affect the uptake of cadmium by crops.

3. In certain countries the environmental concerns are met by legislative measures such as the establishment of cadmium limits in fertilizers.

Selective purchases of low cadmium phosphate and fertilizers

1. Should future demand for low cadmium phosphate exceed actual known resources, prices might rise with an associated impact on fertilizer prices, thus providing some encouragement for the investment in decadmiation processes.

2. To satisfy demand, technological processes for decadmiation would be essential.

3. Other sources of cadmium in fertilizer production, such as that derived from cadmium contaminated sulphuric acid, might need to be taken into account.

Technology for cadmium removal from rock phosphate/phosphoric acid

Various decadmiation processes currently available for food/feed phosphates or under development are summarised in the appendix to this report. Time scales indicated in the table take into account technical aspects only and not market or other economic constraints.

Obstacles to further development

Further investments in research and development and the implementation of commercial scale production are required before long-term solutions can be achieved. However, under current market conditions, the provision of private funding for further technological development is unlikely. Continuation of international funding (such as that previously provided by the EU) is therefore required.

The exchange of information and knowledge on technical innovation will also be an important factor in fostering the commercial application of decadmiation processes. This needs to include a continuing dialogue between all stakeholders in the problem.

Conclusions

An increasing demand for lower cadmium fertilizers might mean that in the longer term existing resources would not meet the need.

1. Additional demand might be met only by the development of decadmiation processes.

2. On the basis of existing knowledge, some technologies are available but they require further development.

3. Funding of this development may not be available from private sources; therefore public (national or international) support could be essential.

4. Based on current market prices for phosphoric acid (approximately \$400/tonne P_2O_5), the estimated additional cost for decadmiation of acid should not be greater than 10 per cent.

OVERVIEW OF DECADMIATION PROCESSES

	Stage of development				Achievable level in mg Cadmium/kg P	Residue * [1]	Decadmiation cost [2] US Dollar/t P_2O_5	Process-operator
	Lab bench	Micropilot	Pilot	Pilot plant				
ROCK thermal treatment	Yes	Yes	< 5 years	10 years? *	~ 60	Solid waste 0.1 - 0.2% Cadmium	30 (basis 350.000 t/y)	Supplier
ACID [3] Cocrystallisation $CaSO_4$	Yes	Yes	< 5 years	10 years? *	< 10	Solid waste 0.1- .2% cadmium Cadmium metal (95%)	6 9	Supplier
Precipitation with sulphide	Yes	Yes	Yes	< 5 years Already used for feed-phosphate 100.000 t P_2O_5/y	< 30 (actual level depends on amount of sulphide used)	Solid waste [4] < 1% cadmium	30	Supplier or user
Solvent extraction	Yes	Yes	Yes	< 5 years Already used for food/feed-phosphate 60.000 t P_2O_5/y	< 5	Filter cake/ concentrate [4] (30-60% cadmium)	32	Supplier or user
Ion exchange	Yes	Yes	Yes	< 5 years *	< 5	Solid waste [4] (5-10% Cadmium)	30	Supplier or user

(*) Further development needed.
(1) In general terms, the lower the level of cadmium in the residue, the greater the volume of that residue.
(2) Cost includes capital and operation costs and applies to a 150,000 t P_2O_5/y plant (unless otherwise noted).
(3) Several different sources of acid/rock tc be tried out.
(4) Further treatment of the residue to obtain cadmium-metal is possible at a cost.

SESSION B

IMPLICATIONS OF MEASURES TO REDUCE THE LEVELS OF CADMIUM IN FERTILIZERS

The Existing Instruments for Environmental, Technical and Financial Co-operation Regarding Africa and the Middle East

Marianne Laanatza

Sweden

Market conditions

The importance of and dependence on rock phosphate imports from Africa and the Middle East are underlined in the consultant report on *Cadmium in Fertilizers* presented at this workshop. The problems related to cadmium in fertilizers have been presented, as well as how individual OECD countries have handled these problems. Some have introduced their own national regulations or voluntary agreements. The restrictions are more extensive among some of the EU Member States than among OECD countries in the Pacific region.

Taking notice of the tendency within the OECD group to introduce regulations regarding the maximum content of cadmium in fertilizers, more focus has been put on the need for larger quantities of phosphate with low cadmium content. This need is expected to increase in the future.

The possibility and ability to offer customers exactly those qualities demanded varies from country to country, depending on very different prerequisites. The conditions in which the cadmium content of the phosphates is high, as in Togo and Senegal, can hardly be compared with the situation in which there is a lower cadmium content, in Morocco and Jordan.

These different conditions have already created a market segmentation. It has became an important instrument for competition to offer so-called "tailed" products. A further development of these trends will create a more and more complicated market structure.

The lack of the demanded quantities of phosphate with low cadmium content may lead to higher prices for such phosphate, and even to possible market disturbances. Those developing countries which can offer the demanded qualities may gain, while those which cannot compete may lose. For those developing countries which are very dependent on export incomes from their phosphate production, uncertain market conditions may have a negative impact on their overall prospects.

Altogether there are many reasons for the OECD countries to take such perspectives into consideration. Developing countries which are particularly dependent on phosphate exports include Morocco, Tunisia, Jordan, Senegal, Togo and Nauru.

Financial instruments and sources

Up to now, phosphate production in the above mentioned developing countries has received financing from both public and private sources. The International Monetary Fund (IMF) and the International Development Association (IDA) are stressing the importance of including the phosphate industry in the liberalization programmes of the economies concerned, as a part of individual structural adjustment programmes.

Expectations regarding further engagement in the phosphate production sector in developing countries from the side of the World Bank may increase, since the Bank recently presented its new priorities. Environmental aspects and projects are particularly stressed. The EU policy and attitude towards structural adjustment programmes, and environmental aspects and priorities, are the same as the World Bank's. Possibilities for co-financing, or at least parallel financing, by the World Bank and the EU have probably increased as a result of the Bank's new priorities.

For historical and geographical reasons, engagement in and financial support to most of the African and Mediterranean countries have been, and still are, more extensive from the EU Member States than from other countries within the OECD. Therefore, most of the examples are from EU activities.

As presented in the consultant report *Cadmium in Fertilizers*, since the 1970s the EU has had bilateral agreements with several Mediterranean non-member countries, including Morocco, Tunisia and Jordan, covering preferential trade and financial and technical co-operation. Furthermore, the EU has additional economic resources for environmental projects in the Mediterranean region which are available to the same non-member countries. The total amount of aid is 230 M ECU and the EIB credits are 1800 M ECU for a period of five years ending in 1996.

Recently a new agreement on free trade was signed between Tunisia and the EU, and the EU is also negotiating with both Jordan and Morocco to sign such so-called "association agreements". The EU has formulated a new Mediterranean policy, including a network of association agreements and a special aid programme covering 4.67 billion ECU.

Within the new EU policy, including its aid programme, both structural adjustment and environmental aspects are stressed. Thus the possibilities to receive assistance from the EU for projects in the phosphate production sector regarding cadmium reduction seem to have increased.

Sub-Saharan African countries like Senegal and Togo, and a number of former European colonies in the Caribbean and Pacific regions like Nauru, are covered by the EU's Lomé conventions (presented in the consultant report). The current convention, which includes preferential trade conditions, aid and technical assistance from the EU, also has a special instrument for the mining sector, the SYSMIN, to stabilize price variations from one year to the other. A special European Developing Fund (EDF) covers the aid programme included in the Convention. The current fund, the Seventh, amounts to 10.8 billion ECU and is tied to 1.2 billion ECU in risk capital from the EIB. Economic resources for the following

Eighth European Developing Fund have been fixed at 13.132 billion ECU and are tied up with 1.693 billion ECU in risk capital from the EIB.

Financing of projects, which include both mining and environmental aspects through the EDF, are included in the target. The position gets even stronger when structural adjustment aspects are also included in the project.

Possible combinations of market qualifications and financial possibilities

The strategy stressed by the IMF, the World Bank and the IDA, as well as other financial institutions for developing countries, particularly in the EU's financial instruments including the EIB and EDF, seems to be more co-ordinated than ever. All are stressing the importance of structural adjustment programmes, including environmental aspects.

This approach, combined with a stricter attitude towards cadmium content in fertilizers in several OECD countries, might pave the way for more resources being allocated to projects within the phosphate sector in the developing countries concerned.

Those cases known concerning direct investments in projects in Africa and the Mediterranean countries to reduce the cadmium content in phosphate are all linked to the EU.

The information, positions and attitudes presented in the report *Fertilizer Industry of the European Union*[3] show that the EU is highly aware of the European fertilizer industry's great dependence on imports of rock phosphate from North Africa, etc. and that available quantities of desirable low cadmium-containing rock are insufficient in a global perspective. This has placed the emphasis on developing effective and viable cadmium removal processes for use during the manufacture of fertilizers. Whilst no such commercially viable process exists, an EU-sponsored programme is in progress in North Africa to develop cadmium extraction processes which could be employed in the production of phosphoric acid.

Resources from the EIB have already been used for Mediterranean environmental projects. It is important to stress that the results of the EU-sponsored cadmium extraction process developed in Morocco will be available for all the Mediterranean countries concerned (information from DG-I, the European Commission, 1995).

The current situation is such that a pilot project is planned for Morocco, with costs calculated to be around 7 M ECU. The European Commission is examining its eventual financial support. There are different possibilities, both within the EU's current financial instruments for environmental projects in the Mediterranean region and with future resources covered by the planned MEDA regulation. (Personal communication, European Commission, 1995.)

---

[3] Published by the European Fertilizer Manufacturers' Association with the co-operation of the Directorate General for External Economic Relations (DG-I) and the Directorate General for Industry (DG-III) of the European Commission, 1994.

As mentioned in *Cadmium in Fertilizers*, the EU has sponsored programmes in two African countries, Senegal and Togo. The aim was to develop methods for reducing or eliminating cadmium from rock phosphate. The methods used turned out to be uneconomical. The EU Commission is examining the possibility of using SYSMIN[4] resources to finance tests of Senegalese phosphate samples in Morocco within its test programme for cadmium reduction. Up to 150,000 ECU could eventually be allocated for this purpose. Senegal's eventual participation in the above mentioned pilot plant in Morocco is not yet decided. (Personal communication, the European Commission, DG-VIII, 1995).

To sum up the positions and possibilities for future investments in large scale plants for reducing cadmium in phosphate, the picture is not clear, although the efforts to combine structural adjustments and environmental aspects as a platform for decision making in different international financial institutions, particularly the IMF, the World Bank and IDA, are a step forwards. The projects regarding cadmium which have already been undertaken or assisted by the EU have paved the way for new initiatives. The above mentioned pilot plant project in Morocco is another important step forwards. A positive outcome of this proposed project will, however, be necessary to plan any large scale plant.

---

[4] SYSMIN is a facility specifically designed to provide financial support in the form of grants to ACP countries (covered by the Lomé Convention) encountering difficulties in their mining sector.

References

EFMA, DG-I and DG-III of the European Commission, 1994. The Fertilizer Industry of the European Union, Brussels, Belgium.

EU Commission, DG-I and DG-VIII, 1995. Information from meetings in the Commission, July 1995, Brussels, Belgium

D.I. Gregory, 1992. "Global Structures of the Phosphate Fertilizer Industry." IDFC, Proceedings of an International Workshop, Tampa, Florida.

IMF, 1993-94. Trade Statistics, Yearbooks, International Monetary Fund, Washington, D.C.

Jackson, T. and A. MacGillivray, 1995. "Accounting for Cadmium; Tracking environmental emissions of cadmium from the global economy," chapter 10, in Emissions from fertilizers production and use, Stockholm Environmental Institute, Stockholm, Sweden.

JPMC, 1994. Publication from Jordan Phosphate Mines Co Ltd, Amman, Jordan.

JMPC, 1995. Complementary information from meeting at the Jordan Phosphate Mines Co Ltd, Amman Jordan.

UNCTAD, 1993-94. Statistics, Yearbooks, United Nations Conference on Trade and Development, Geneva, Switzerland, and New York.

World Bank, 1993-95. Yearbooks and The World Bank Atlases, The World Bank, Washington, D.C.

The Importance of the Phosphate Sector to the Economy of Senegal: Experience of ICS in the Reduction of the Cadmium Content in Phosphoric Acid

I. Kotlarevsky and D. Fam
Senegal

Introduction

Chemical Industries of Senegal (ICS) started producing phosphoric acid and phosphatic fertilizer from Taïba phosphate rock in 1984. The major part of this production is exported.

The phosphate sector includes three companies whose aggregate turnover is more than $US 200 million. It is the most dynamic factor of Senegal's industrial activity. With more than 22 per cent of the country's total exports, it is the first source of foreign currency.

The various legislation established by some European countries in connexion with the cadmium content of phosphatic fertilizers practically excluded deliveries of fertilizers from Senegal to Europe.

ICS are developing a research programme to establish a reliable and economical process for the reduction of the cadmium content of phosphoric acid.

This work was financed by the World Bank and by the EU through its SYSMIN programme. Two processes have been particularly explored:

- precipitation and separation with organic reagents;
- precipitation and separation by co-crystallization in anhydrite.

The results of this research work are presented hereafter.

The phosphate sector in Senegal

The activity of the phosphate sector covers four main products:

- tricalcium - phosphate rock;
- aluminium - calcium phosphate;
- phosphoric acid;
- complex fertilizers in various grades.

There are three companies involved:

- CSPT, with a capacity of approximately 2 million tonnes of tricalcium phosphate, of which .9 million is locally transformed into acid by ICS;

- SSPT, with a capacity of .1 million tonnes of rock for export;

- ICS, with a capacity of .33 million tonnes of phosphoric acid (P_2O_5) and .2 million tonnes of fertilizers.

One major factor of importance in this sector is manpower. Directly, there are around 2500 employees and roughly 2500 daily workers. Indirect jobs represent around 3000 persons. Knowing that in Senegal one worker supports ten persons, this means that the sector supports about 80,000 citizens. Wages represent around $US 40 million per year.

Local expenses including spare parts and services amount to $US 50 million per year.

Various important sectors of the national economy are concerned by the phosphatic activity:

- electrical power - $US 30 million (this is 25 per cent of the turnover of the power company;

- water supply - 20 per cent of the activity of the national company;

- railways - 80 per cent of its activity;

- Port of Dakar - 50 per cent of its activity.

Cadmium removal from phosphoric acid

The removal of cadmium from phosphate rock has been studied and abandoned because of the difficulties of high-temperature calcination.

In 1986, ICS started working on the reduction of the cadmium content of phosphoric acid.[i] The first studies were performed in co-operation with European laboratories, which strongly contributed to the development of a process by precipitation.

In order to widen the field of experiments, ICS took advantage of a study financed by the EU[ii] considering the technico-economic feasibility of all known processes.

Other tests were performed in 1991 with the OTP process (Togo).

In all these tests the target was to obtain a cadmium-free acid containing less than 10-20 ppm.

Cadmium removal with an organic reagent

Organic compounds based on di-thio-phosphate are common in the mining industry as collectors for froth flotation. These compounds have, *inter alia*, the capacity to react with several heavy metal ions in aqueous solution, producing a solid precipitate. The fixation of an alkyl radical allows utilization of the same reagent in strong acid solutions. In particular, di-methyl-amine-di-thio-phosphate can remove and precipitate cadmium in phosphoric acid concentrated at 54 per cent P_2O_5. This precipitation is relatively stable and can be separated by filtration or centrifugation.

The objective of ICS work is to develop a technico-economic process aiming at reducing the cadmium content in phosphoric acid from 100-200 ppm to 10-20 ppm.

The results of ICS studies are presented below:

Analysis of the phosphoric acid

These studies were made using an atomic absorption spectrophotometer.

Table 1 indicates the concentration of each metal:

Table 1

Metal concentrations

Metal	Cd	Cu	Zn	Ni	Mn	Fe
ppm	120	74	882	74	280	1000

Reagent

The reagent was the di-thio-phosphate type with the generic formula:

$(RO)_2PSSNa$

R being an alkyl group.

Performance of laboratory tests

Test were performed with 100 millilitre samples of phosphoric acid. These samples were placed in a round-bottomed glass with holes to allow mixing, addition and samples up-taking. The reagent is kept in a water bath, thermally controlled.

Results and analysis

Table 2 gives the test results, aimed at determining the minimum quantity of reagent needed. Concentrations are also indicate in this table for each test in function of time, in order to evaluate the precipitate stability.

All the tests were performed at 40°C. Results show that final cadmium content below the 10 ppm level is attainable with the addition of twice the theoretical quantity of reagent needed for removing all cadmium. The reagent quantity is 7.6 kg per tonne of P_2O_5.

The following chemical reaction allows the calculation of the theoretical need:

$$Cd^2 + 2(RO)_2PSS\ Na \rightarrow [(RO)_2PSS]_2Cd + 2Na \qquad (1)$$

Table 2 also contains data on the residual content of copper and zinc for each dose of reagent. It can be seen that the reagent has the following order of selectivity: first copper, second cadmium, third zinc.

The utilization of the minimum quantity of reagent in order to obtain the target of less than 10-20 ppm of cadmium is therefore performed in two steps.

If one agrees that a formula similar to (1) can be used for copper, then:

$$Cu2 + 2(RO)2PSS\ Na \rightarrow [(RO)2PSS]2Cu + 2Na \qquad (2)$$

and one can calculate the theoretical needs of reagent:

- for 74 ppm of copper, the needed reagent is 1230 mg per litre of phosphoric acid at 54 per cent P_2O_5;
- for 120 ppm of cadmium, the need is of 1208 mg.

These values correspond exactly to 7.6 kg of reagent per tonne of P_2O_5 (acid). This value represents the minimum quantity required for removing cadmium and copper. This consumption of reagent depends therefore on the total content of cadmium. Copper in the acid and copper alone takes the excess quantity of reagent.

Other impurities in solution in the acid, zinc more particularly, do not seem to play a role in this consumption, provided the dose of reagent is kept equal to the theoretical one as dictated by the content of copper and cadmium.

Table 2

Variations in residual cadmium content

Reagent dose		Contact time	Metal concentration		
Volume	Theoretical dose	min	Cd ppm	Cu ppm	Zn ppm
0.329	1	5	76	19	700
	1	30	77	0.1	
	1	60	77	<0.1	
0.493	1.5	5	42	4	
	1.5	30	40	0.2	
	1.5	60	40	<0.2	
0.657	2	5		<0.1	
	2	30	23	<0.1	806
	2	60	14	<0.1	811
	2	120	1	<0.1	683
	2	270	1	<0.1	789
	2	48 HR	17	2	850
0.724	2.2	5	3	<0.1	
	2.2	30	2	<0.1	
	2.2	60	2	<0.1	
0.79	2.4	5	24	<0.1	
	2.4	10	216	<0.1	
	2.4	15	13	<0.1	
	2.4	30	18	<0.1	
	2.4	45	6	<0.1	
	2.4	60	6	<0.1	
	2.4	120	3	<0.1	
	2.4	240	3	<0.1	
	2.4	480	2	<0.1	
	2.4	24 HR	3	<0.1	
0.987	3	5	1	<0.1	722
1.32	4	5	0.4	<0.1	689
1.645	5	5	0.4	<0.1	656

The variations in residual cadmium content indicated in Table 2 (for identical conditions of reagent - temperature - time) seem to be connected with the mixing efficiency. This is confirmed by the low speed of cadmium precipitation.

In summary, one can say that the D.M.D.T.P. allows an efficient precipitation of the cadmium contained in the phosphoric acid produced by ICS, when added at twice the theoretical dose. This precipitate is stable and can be separated by filtration or centrifugation.

Cadmium removal by anhydrite co-crystallization

Tests were performed in a semi-industrial pilot plant by Duetag Company with the so-called OTP process on ICS phosphoric concentrated at 53 per cent P_2O_5.[iii]

The process consists in a co-cristallization of cadmium and calcium in anhydrite.

Table 3

Analysis of the phosphoric acid (per cent)

Cd	P_2O_5	Fe_2O_3	Al_2O_3
107 ppm	53	1.37	0.93

Pilot plant description

The pilot plant consists of a main reactor where the following are introduced:

- phosphoric acid heated at the reaction temperature;

- sulphuric acid;

- monocalcium phosphate.

The reaction is performed during a maximum of two hours. Then the slurry is filtered on a vacuum filter or press filter operating under a pressure of 5 bars (see Figure 1).

Results

Results of the cadmium removal tests are indicated in Figures 2 to 5. They show the monocalcium phosphate solution.

Figure 1

Slurry filtration

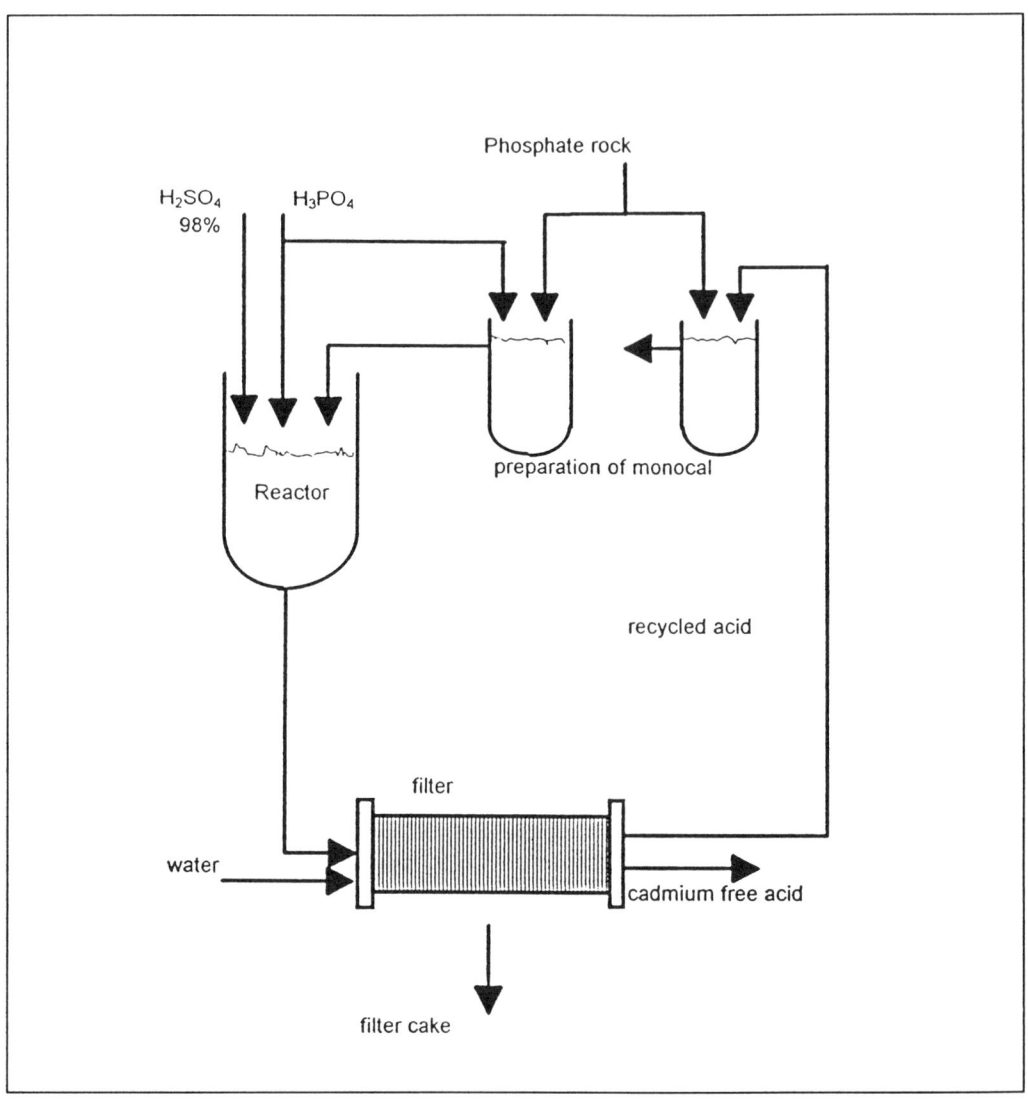

The same graphs show the evolution of sulphuric acid content (in per cent H_2SO_4) decreasing with the addition of calcium ions, as well as P_2O_5 concentration, which varies also during the reaction due to:

- addition of monocalcium phosphate;
- water evaporation.

Table 4 shows the cadmium removal efficiency during the eight different tests performed:

Table 4

Cadmium removal efficiency

Test No.	Conditions	ppm Cd	Cd reduction (%)
1	Silicate added	12	88.79
2	"White" test	22	79.44
3	P_2O_5 concentration effect	17	84.11
4	Silicate added	9	91.59
5	Organic effect	12	88.79
6	Settled acid	18	83.18
7	Redox effect	12	88.79
8	Gypsum effect	10	90.65

One can see that cadmium removal is:

- 80-84 per cent efficient without any additive;

- 98-92 per cent efficient with 2 per cent silicate addition.

The destruction of organics and the redox effect have not brought about any significant improvement.

Sludges filtration

The filtration of the precipitate formed by the reaction is performed under vacuum or with pressure. The vacuum system has a limited strength (around 500 mm mercury).

The pressure system allows high strength up to 16 bars, but the tests were limited to 5 bars.

Results of filtration tests are indicated below:

- *Vacuum filtration*

Vacuum filtration was performed at 90-100°C with 500 mm mercury vacuum and three washing steps. Table 5 indicates the main results.

Table 5

Results of vacuum filtration

Test No.	Filterability Cycle 2 mm in tonnes P_2O_5/m^2 per day	Filterability Cycle 4 mm in tonnes P_2O_5/m^2 per day
1	13.3	9.0
2	19.3	13.7
3	16.4	11.4
4	13.8	9.8
5	16.1	11.4
6	20.4	14.5
7	11.7	8.2
8	11.7	8.2

- *Pressure filtration*

Pressure filtration was performed at five bars - 100°C, 130 kg of slurry per m². The filter cake had 10-12 mm thickness. The main results are:

− Test No. 5: 9.3 tonnes P_2O_5/m^2 per day - 9 min. cycle - 110 kg/m² load.

− Test No. 7: 10.3 tonnes P_2O_5/m^2 per day - 10 min. cycle - 130 kg/m² load.

In summary, one can say that this process allows cadmium reduction in ICS acid from 107 to 12 ppm. These values were obtained with the use of 2 kg of silicates per tonne of acid; without this use, the cadmium level is 15 ppm. Filterability of the precipitate obtained with pressure filter for vacuum filter is satisfactory.

Conclusion

The cadmium content of the Taïba phosphate rock is a great handicap for good development of the Senegalese phosphate sector.

The development of the present studies can allow the sector to survive if the economic constraints are removed.

The technical results are, however, encouraging.

Optimization of these studies and co-operation with various research centres will allow Senegal to remain competitive in the world phosphates market.

Figure 2

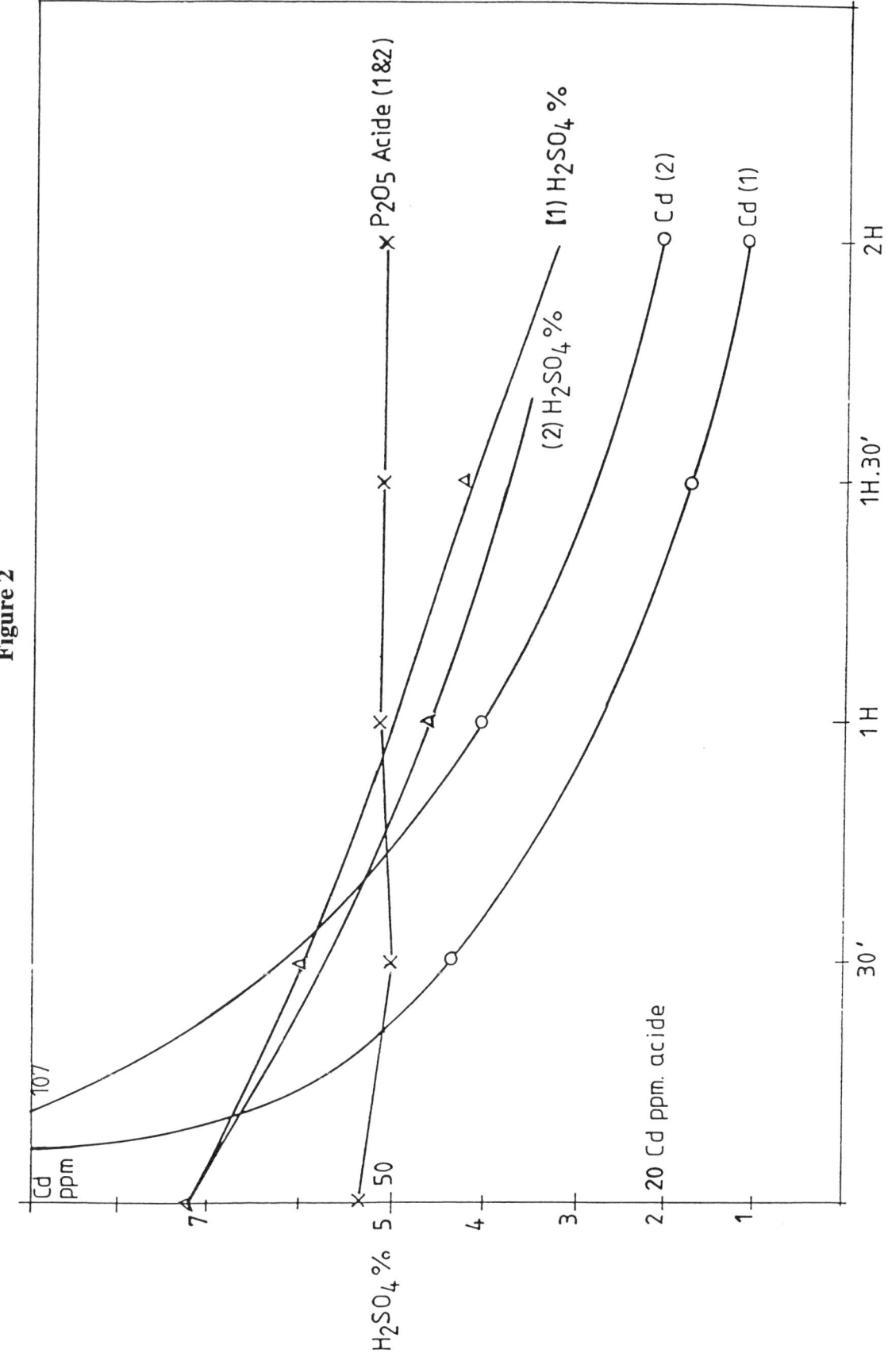

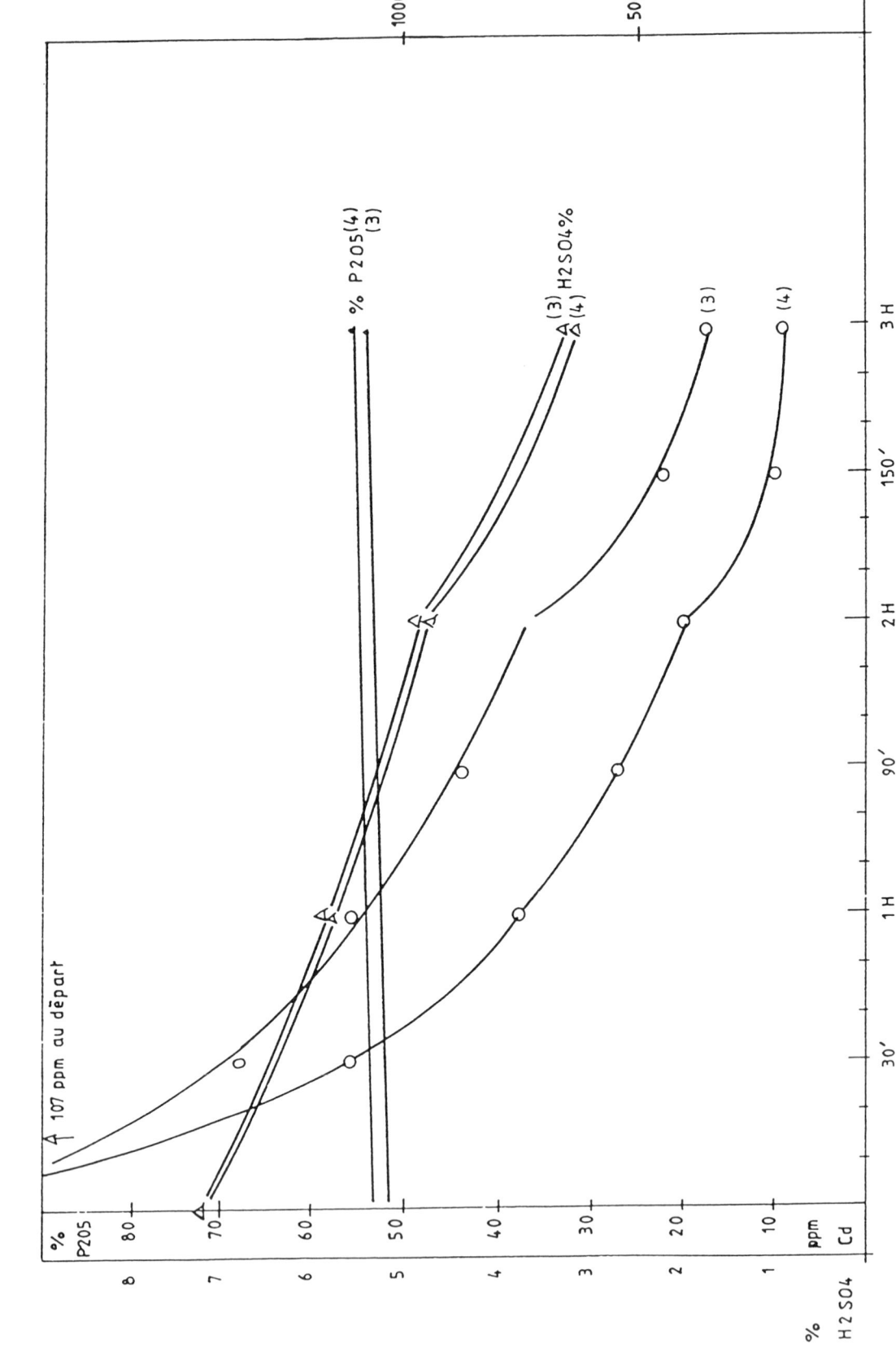

Figure 3

Figure 4

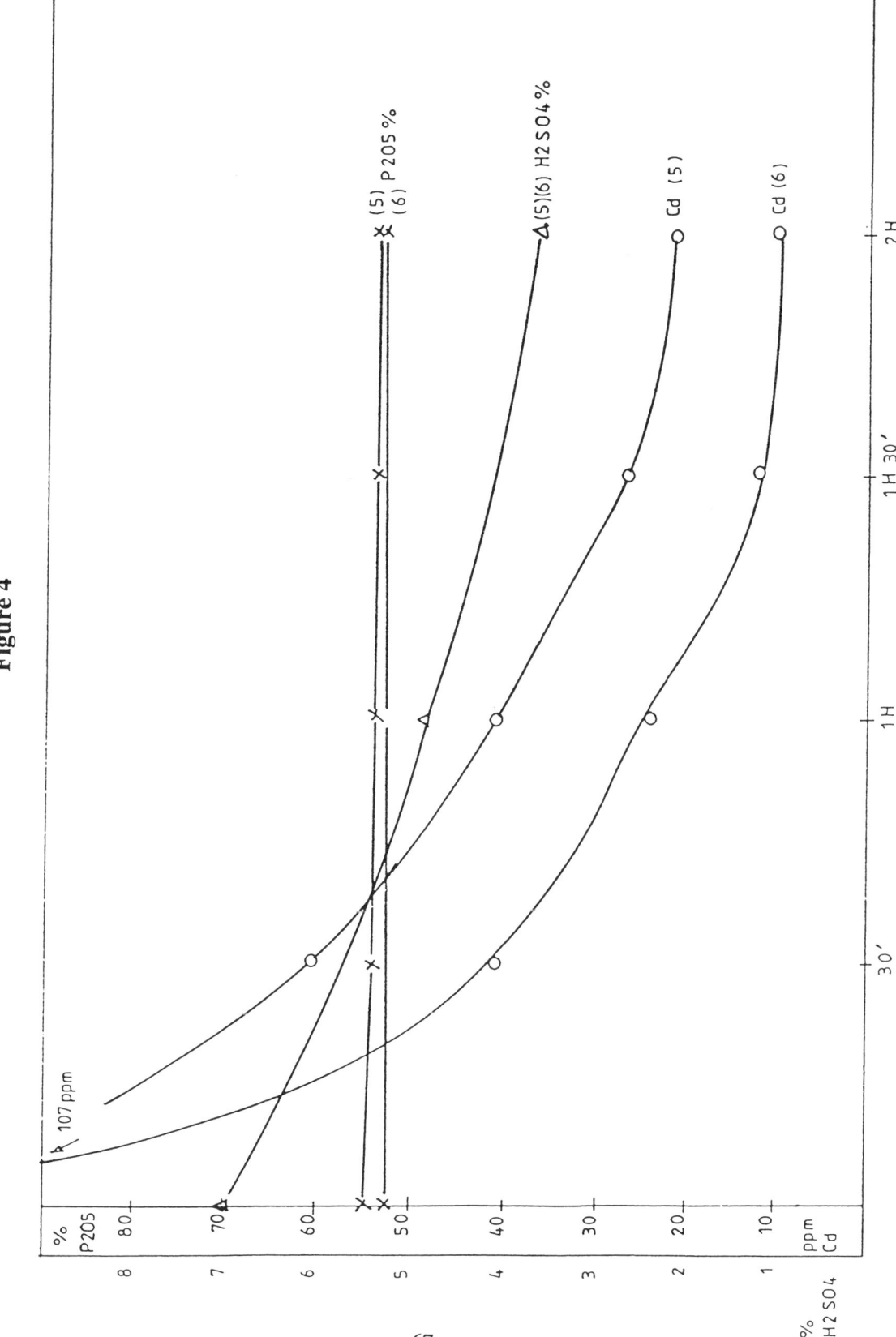

Figure 5

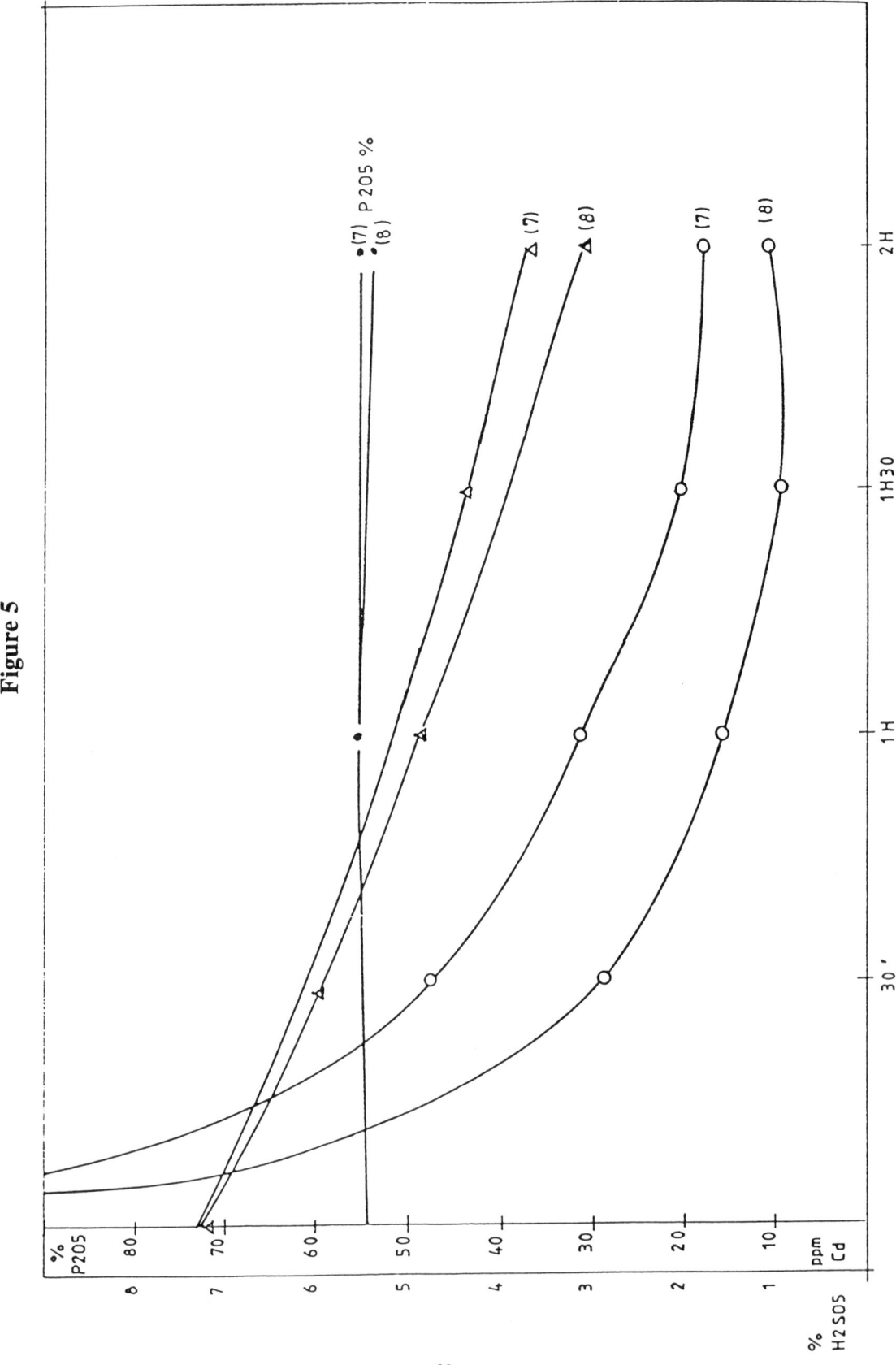

Notes

1. Adam, M.R., Cadmium Removal from Senegalese Phosphoric Acid by Precipitation - CR 3813 (August 1993).

2. Davister, A., Cadmium Removal Process Survey (1992).

3. Duetag, Cadmium Removal from Senegalese Phosphoric Acid by the O.T.P. Process - (October 1992).

Agronomic Implications of Restricting Cadmium Content of Phosphate Rock

T.L. Roberts and M.D. Stauffer

Potash and Phosphate Institute of Canada

Growing health concerns about cadmium have prompted the world community to investigate ways of reducing cadmium content of foods. One option is to reduce or restrict the cadmium that occurs naturally in phosphorus fertilizers.

Phosphate rock is widely distributed throughout the world as marine phosphorites, apatite-rich igneous rock, and modern and ancient guano. This vital resource is the only economic source of phosphorous for production of phosphate fertilizers.

Production and reserves of phosphate rock

The extent of world phosphate rock reserves is rather uncertain and often difficult to determine. Current United States Bureau of Mines estimates of reserves and reserve base are shown in Table 1. Morocco dominates with over 50 per cent of world reserves. South Africa, the United States and China follow, accounting for another 35 per cent of listed reserves.

The above does not include any estimates for reserves in the African countries of Senegal, Togo and Tunisia, or the former USSR countries of Russia and Kazakhstan. However, reserves in Senegal, Togo and Tunisia have been estimated at 85, 130 and 460 million tonnes, respectively (Savage 1987) and reserves in the former USSR at 1,200 million tonnes, or 10 per cent of the world total (Stowasser 1991). About 60 per cent of production in the former USSR was from igneous rock and 40 per cent from sedimentary rock. Profitable reserves may increase as new deposits are discovered and as technology improves. Reserve estimates can also change with changing economic and environmental conditions.

World production of phosphate rock ranged from 121 to 141 million tonnes between 1992 and 1994 (Table 1). Average production (129 million tonnes) was down over 30 million tonnes from its peak in the late 1980's due to the near collapse of output from the former USSR. Eleven countries produce over 90 per cent of the world's phosphate rock. China, Morocco (including the Western Sahara), Russia and the United States account for 70 per cent of world production.

Table 1

World production of phosphate rock

Countries	Reserves	Reserve base	Production		
			1992	1993	1994
	million tonnes		thousand tonnes		
China	210	210	23,000	24,000	24,000
Israel	180	180	3,600	3,590	3,600
Jordan	90	570	4,300	3,570	3,500
Kazakhstan	---	100	7,000	4,000	3,000
Morocco and Western Sahara	5,900	21,440	19,184	18,300	18,000
Russia	---	1,000	11,500	9,400	8,000
Senegal	---	160	2,280	1,670	1,600
Republic of South Africa	2,500	2,500	3,080	2,470	2,880
Togo	---	60	2,080	1,750	1,800
Tunisia	---	270	6,400	5,500	6,500
United States	1,200	4,440	47,000	35,500	41,100
Other	1,300	2,900	11,660	11,250	11,020
Total (rounded)	11,000	34,000	141,000	121,000	124,000

Reserve and reserve base cost less than $40 and $100 per tonne, respectively. Cost includes capital, operating expenses, taxes, royalties, 15 per cent return on investment. FOB mine.
Source: Morse (1992), US Bureau of Mines (1995).

Sedimentary deposits dominate world reserves and production. They accounted for about 90 per cent of world phosphate rock production in 1994. The remaining production was from igneous deposits, primarily in Russia and South Africa. Production from igneous phosphate rock deposits has declined steadily over the past five years from approximately 18 per cent of total production in 1990 to about 10 per cent in 1994. This decline is largely due to a decrease in production from the former USSR.

Cadmium content of phosphate rock

Phosphate rock naturally contains, in addition to phosphorous, trace amounts of cadmium and other heavy metals. Amounts vary widely depending on the source of rock. Table 2 shows typical cadmium concentrations of various sources of phosphate rock. Concentrations range from 3 to 150 ppm in sedimentary rock and are usually less than 2 ppm in igneous rock.

Table 2

Typical cadmium concentrations of phosphate rock deposits

	Mean	Range
Sedimentary deposits	ppm	
China	2	1-4
Israel		
Arad	14	12-17
Zin	31	20-40
Undifferentiated	24	20-28
Jordan	5	3-12
Morocco		
Khourbiga		
Youssoufia	15	4-19
Undifferentiated	26	10-45
Western Sahara	38	32-43
Senegal	87	60-115
Togo	58	48-67
Tunisia	40	30-56
USA		
Florida Central	9	3-20
Florida North	6	3-10
North Carolina	38	20-51
Western US	92	40-150
Igneous deposits		
Republic of South Africa	1	<2
Former USSR	1	<2

Source: IFDC and TVA, unpublished data, and others.

Cadmium is fairly evenly distributed throughout the rock and carries through the beneficiation and acidulation processes during fertilizer manufacture. Studies in the United States have shown that the distribution of cadmium is about one-third to gypsum and two-thirds to filter-grade acid during processing, depending on the source of rock (Wakefield 1980). Concentration of cadmium in the final fertilizer product is approximately equal to the sum of cadmium in the starting materials (i.e. phosphoric acid or rock).

Technologies are available, and under development, which can reduce cadmium content of phosphate rock and phosphoric acid during processing. Numerous techniques have been investigated. These include: solvent extraction, ion exchange, precipitation, electrostatic precipitation, high-temperature calcination and others (Anonymous 1987). However, the technology is complicated and costly. Furthermore, it may not be effective, or accessible, for all sources and producers of high cadmium rock due to the varied properties of the rock and the different manufacturing processes.

Agronomic implications of restricting high cadmium phosphate

Implementation of technologies to reduce cadmium content of phosphate rock would undoubtedly lead to higher production costs and higher input costs to end-users. And, restricting use of high cadmium phosphate rock (i.e. > 20 ppm cadmium) could potentially eliminate production from about 60 per cent of present reserves. Either option could reduce the use of phosphorous fertilizers and impact the use of nitrogen and potassium fertilizers. Any time fertilizer use is altered, crop and fibre production is affected.

Food production is directly related to nutrient supply. Any yield increase, above some base level, is dependent on an adequate supply of all essential plant nutrients. Figure 1 illustrates the close relationship between crop production and fertilizer (nutrient) consumption in Canada and China, two countries with diverse agricultural production systems.

Good phosphorous nutrition is vital for producing high yielding crops and optimizing other plant nutrients, especially nitrogen. Decreased phosphorous consumption, due to either higher costs or reduced availability, leads to unbalanced application of plant nutrients. Imbalance of fertilizer use causes losses in yield and crop quality, reduces nutrient use efficiency and depletes plant nutrients from soil reserves.

Application of the right amount and the right ratio of applied nutrients is essential for sustainable production and for proper nutrient utilization. The positive effect of phosphorous fertilization on yield and fertilizer efficiency is illustrated in Figure 2 with data from the Canadian prairies. Nitrogen application tripled wheat yields. But when phosphorous was applied with the nitrogen, an additional 1,100 kg/ha of wheat was produced. Phosphorus fertilization also increased the efficiency of applied nitrogen. At the highest application rates, phosphorous fertilization produced an extra 10 kg of grain per kg of applied nitrogen, resulting in a 37 per cent increase in nitrogen use efficiency.

Such examples of the positive interaction between nitrogen and phosphorous are common and occur throughout the world. Table 3 summarizes data from extensive trials in India conducted over a number of years. All trials were responsive to applied nitrogen, but application of phosphorous with nitrogen increased yields by an additional 22 to 56 per cent over nitrogen alone. And, nitrogen, phosphorous and potassium applied together further increased yields as much as 25 per cent. Correct nutrient balance is needed to produce higher yields and is key to sustaining higher production levels. It is also essential for greater nutrient use efficiency and reduced risk of nitrogen loss to groundwater.

Long-term crop rotation data from the Canadian prairies illustrate how balanced fertilization can prevent the build-up of nitrates in the subsoil. In Saskatchewan, a fallow-wheat-wheat rotation (F-W-W) fertilized with nitrogen alone leached more nitrate-N below the rooting zone than the F-W-W rotation fertilized with both nitrogen and phosphorous (i.e. 286 vs. 191 kg/ha; Figure 3). During the 24-year period, the rotation fertilized with nitrogen and phosphorous received 120 kg/ha more nitrogen than the rotation fertilized with only nitrogen, yet it was better able to utilize the applied nitrogen Balanced fertilization leads to greater crop growth and greater uptake of soil nitrogen, leaving less nitrate-nitrogen available for leaching.

Figure 1

**Grain production and fertilizer consumption in Canada and China
(data in million tonnes)**

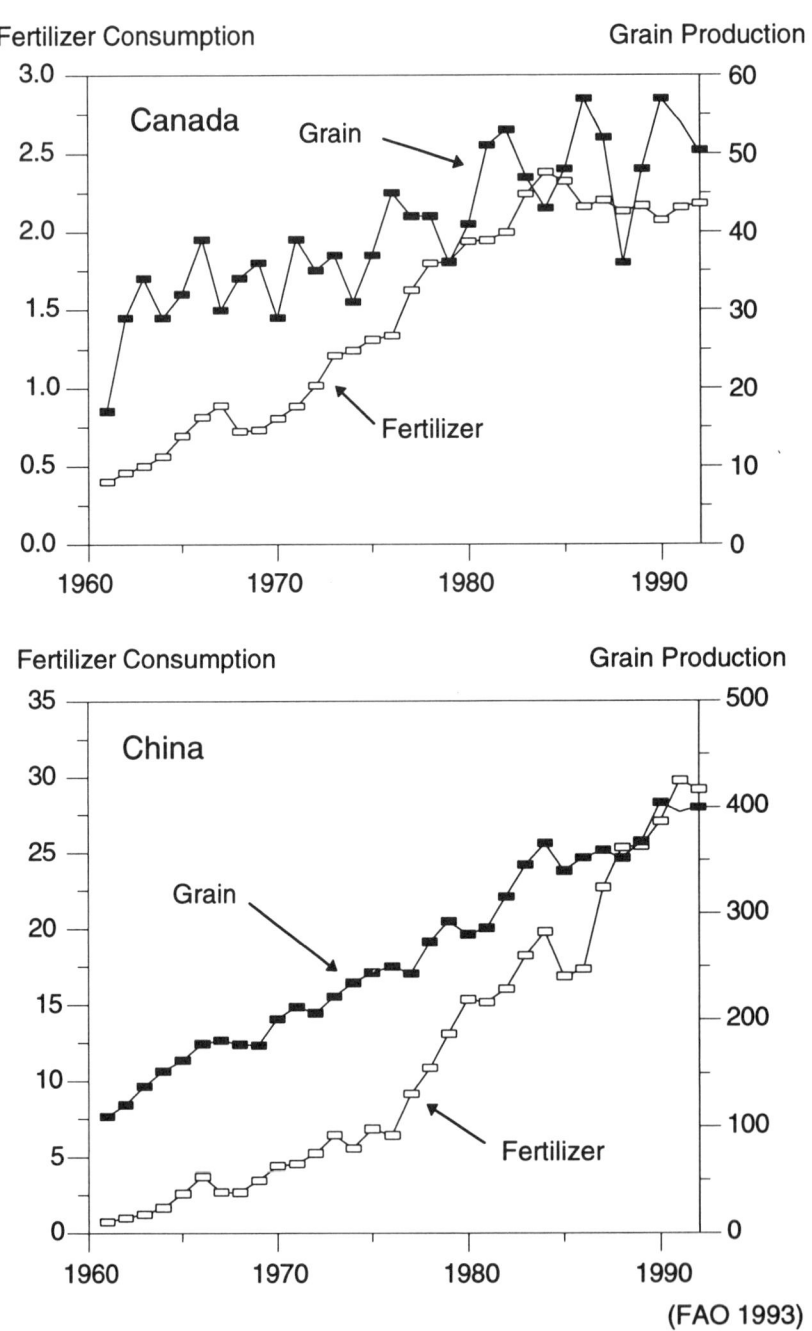

(FAO 1993)

Figure 2

Nitrogen and phosphorus fertilization influence winter wheat yields in the Canadian prairies

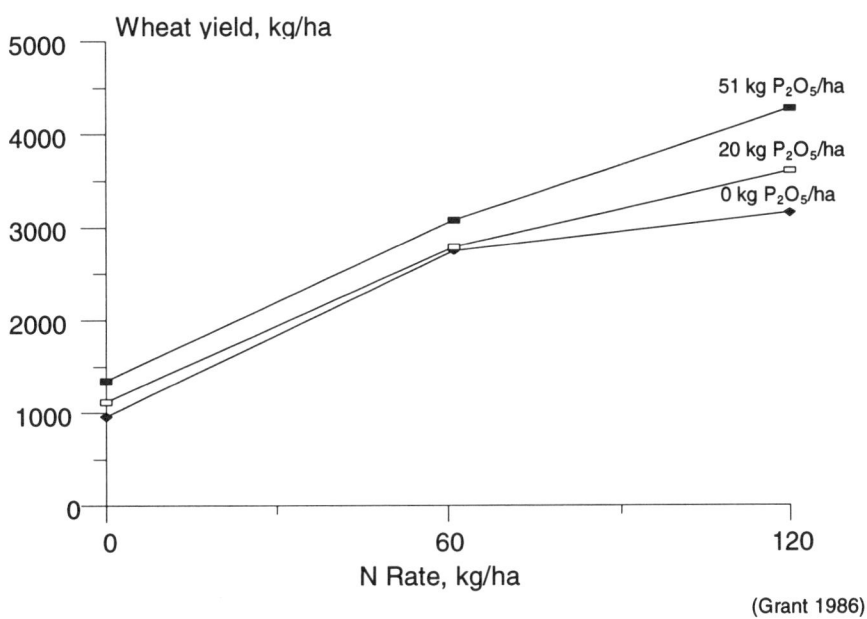

(Grant 1986)

Table 3

Balanced fertilization increases crop yields in India

Crop	Season Condition	Trials	$N-P_2O_5-K_2O$	Yield without fertilizer	Yield response to: N	NP	NPK
			kg/ha	t/ha	%		
Rice	K,U	380	120-60-40	2.42	49	74	99
	K,I	9,634	120-60-60	2.96	27	51	56
	R,I	5,686	120-60-60	3.20	28	51	53
Wheat	R,I	10,133	120-60-60	1.55	59	95	114
Maize	K,U	53	90-60-60	1.23	85	107	129
Millet	K,U	207	90-60-60	0.50	54	110	130
Chickpea	R,U	1,325	20-40-20	0.75	36	59	77

K=Kharif, R=Rabi (seasons); I=irrigated; U=unirrigated
Source: Randhawa and Tandon (1982)

Imbalanced fertilization mines the soil of its most deficient nutrients. Once critical levels are reached, yields fall dramatically, even though large amounts of other nutrients might have been applied. Farmers must pay constant attention to ensure that both correct ratios and correct amounts of nutrients are applied. This is of concern in many regions of the world, both in developed and developing countries. When more nutrients are exported in crops than are returned to the soils through fertilizers or other inputs, soil fertility and productivity are endangered.

Economic implications of restricting high cadmium phosphate

Phosphorus use by farmers is strongly determined by: (i) economics (i.e. fertilizer cost) and (ii) agronomic consideration of crop requirements, and soil content and supply. Often, the former outweighs the latter, especially in developing countries with large numbers of subsistence farmers. For example, in 1992 the Government of India decontrolled the prices of phosphate and potash fertilizers and lowered the set price of urea. This changed the trends in fertilizer use the following year (Beaton et al. 1993). Phosphorus fertilizer consumption decreased by 13 per cent and potassium fertilizer decreased by 35 per cent, whereas nitrogen fertilizer consumption increased by 5 per cent.

Fertilizer prices may not have the same impact on consumption in countries with more developed agriculture, but they do influence fertilizer use. Recent price increases in Canada, due to world markets, have negatively influenced fertilizer consumption trends. Despite generally good growing conditions and phosphorous deficient soils, phosphorous consumption in the Canadian prairies was less than anticipated during the 1994/95 season. In the Province of Alberta, Canada, where high phosphorous demanding crops like canola are grown, about 6 per cent less phosphate was used in 1994/95 than the previous year.

Summary and conclusions

Any measures aimed at reducing or restricting the cadmium content of phosphorous fertilizers would undoubtedly lead to reduced supply, increased costs, and less consumption of phosphorous fertilizers in many areas of the world. If phosphorous consumption declines, soil fertility declines, nutrient balance is upset, nutrient use efficiency is reduced, environmental protection is sacrificed, and crops yield less. Reduced phosphorous consumption would generate less returns at the farm gate and less returns to industries and countries which rely on phosphate rock and phosphorous fertilizer exports.

The social, economic and environmental implications of reduced phosphorous consumption would be far-reaching. World population is projected to approach 6.2 billion people by the year 2000 and 8.3 billion by 2025 (Borlaug and Dowswell 1993). With most of the population increase occurring in developing countries where there is already a food deficit, can developed countries afford to do anything which might jeopardize present and future food production potential?

The cadmium content of phosphorous fertilizers can be reduced or restricted, but at what cost? And, one must question if restricting cadmium content of phosphorous fertilizers would have the desired effect of reducing cadmium content of foods, especially in Canada and other countries where natural background levels of cadmium in the soil are high relative to cadmium additions from phosphorous fertilizers.

Figure 3

Balanced fertilization reduces subsoil nitrates in 24-year fallow-wheat-wheat (F-W-W) rotations in the Canadian prairies

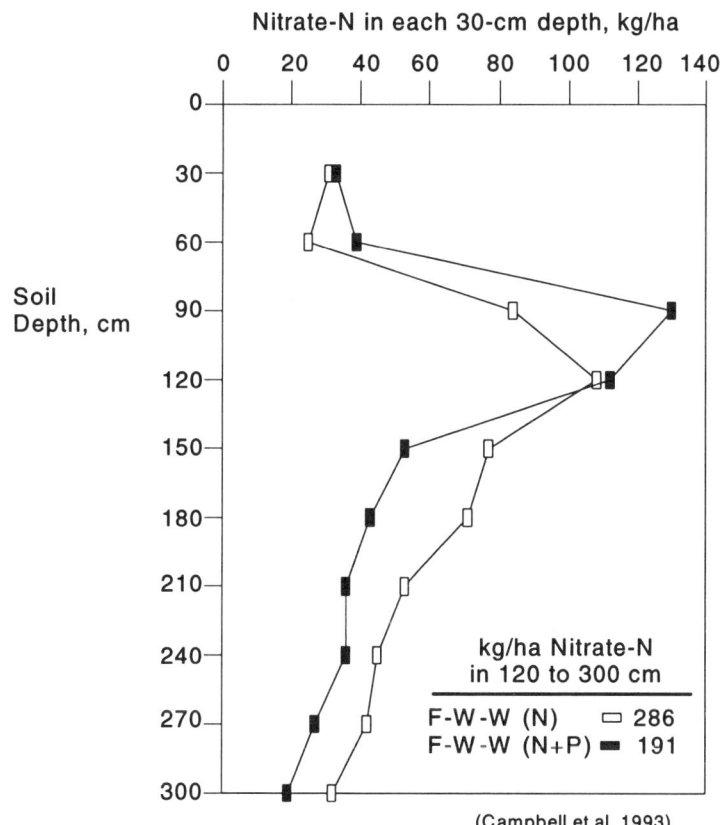

(Campbell et al. 1993)

References

Anonymous. 1987. Cadmium no problem in the EEC and too costly to remove. Phosphate and Potassium, No. 149, p. 40.

Beaton, J.D., G. Dev and E.H. Halstead. 1993. Changing pattern of phosphorous use and its impact on balanced fertilizer use. In FAI Seminar "Emerging Scenario in Fertiliser and Agriculture Global Dimensions," December 6-8, New Delhi.

Borlaug, N.E. and C. R. Dowswell. 1993. Fertilizer: to nourish infertile soil that feeds a fertile population that crowds a fragile world. 61st Annual Conference, International Fertilizer Industry (IFA) Association, May 24-27, New Orleans, Louisiana.

Campbell, C.A., R.P. Zentner, F. Selles and O.O. Akinremi. 1993. Nitrate leaching as influenced by fertilization in the Brown soil zone. Can. J. Soil. Sci. 73:387-397.

FAO. 1993. Computerized Information Series Statistics (Land Use and Population). Food and Agriculture Organization of the United Nations, Rome.

Grant, C.A. 1986. Nitrogen-phosphorus interactions in winter wheat production. Better Crops with Plant Food 70:16-17.

IFDC (International Fertilizer Development Center), Muscle Shoals, Alabama. Unpublished data.

Morse, D.E. 1992. Phosphate rock. U.S. Department of the Interior, Bureau of Mines, 17 p.

Randhawa, N.S. and H.L.S. Tandon. 1982. Fertilizer News 26(2):11-26.

Savage, C. 1987. World survey of phosphate deposits. The British Sulphur Corporation Limited, London, England, 274 p.

Stowasser, W.F. 1991. Phosphate rock. In: Mineral Commodity Summaries, U.S. Department of the Interior, Bureau of Mines, 196 p.

TVA (Tennessee Valley Authority) National Fertilizer Development Center, Muscle Shoals, Alabama. Unpublished data.

U.S. Department of the Interior, Bureau of Mines. 1995. Mineral Industry Surveys. Phosphate Rock. Annual Review 1994.

Wakefield, Z.T. 1980. Distribution of cadmium and selected heavy metals in phosphate fertilizer processing. National Fertilizer Development Center, Muscle Shoals, Alabama. Bulletin Y-159.

Environmental Issues in Relation to Cadmium in Fertilizers

Tayeb Mrabet

Secretary General, World Phosphate Institute (IMPHOS)
Casablanca, Morocco

Abstract

As an institution representing the interests of seven major rock phosphate and fertilizer producing and exporting companies in North and West Africa and in the Middle East, the World Phosphate Institute (IMPHOS) has been a keen observer of environmental issues over the past 20 years. In that regard, the phosphate industry of IMPHOS members, from phosphate rock mining to finished fertilizer production, has made great efforts and investments to comply with environmental standards. Specific actions have been undertaken by IMPHOS to respond to environmental concerns in Western Europe about cadmium in phosphates, namely:

- the implementation of a research programme to explore techniques to reduce the cadmium content of phosphatic fertilizers and to develop technology to remove cadmium during the production of phosphoric acid and fertilizers;

- the review of studies and the collection of research data on cadmium sources and cadmium removal technologies;

- the documentation of international concerns and measures to limit the exposure to fertilizer cadmium via food crops: more particularly, the scrutiny of European national and Community environmental legislation and policy initiatives to regulate maximum cadmium concentrations in phosphate fertilizers.

Without exaggerating the impact of these control measures upon global phosphate fertilizer production and supply, the IMPHOS phosphate industry remains very vulnerable because of the economic and trade constraints on production countries, which rely to a considerable degree on the income derived from the export of phosphates in all forms. IMPHOS is stepping up efforts in the area of research, consultations and dialogues in the search for adequate and widely acceptable solutions to the problems at issue.

Presentation

The recommended application of the "Best Available Techniques" (BAT) in the fertilizer industry for prevention and control of pollution associated with the manufacturing of its products is one of the latest attempts by that industry to achieve the most cost-effective environmental solution, something which deserves wide recognition.

The phosphate fertilizer industry in particular has a long, good record of achievements. It has contributed for more than a century to the welfare of mankind, its social and economic well-being, and the changes introduced in its life style.

As the urban population expanded rapidly with the industrial revolution in Europe, farmers strived to increase crop production. The manufacture of superphosphate from the early 1840s provided them with a cheap source of phosphorus in a plant available form. The result is a very gradual build-up of soil fertility over the following hundred years.

The increase in crop productivity per unit area of land since the 1950s is truly remarkable, especially when viewed against the background of almost unchanging yields in the previous hundred years. It has been possible to feed many more people than could have been supported using the agricultural practices of the 1930s 1940s.

There are many examples in ancient history of the rise and fall of great civilizations which perished when their agricultural base could no longer support them. If agriculture is to feed an ever increasing population worldwide, fertilizers in whatever form are essential to apply to inherently infertile soils so that the growing population can continue to be fed in a sustainable way.

One hundred and fifty years since Justis Liebeg published his theory on plant nutrition, and a century since sedimentary phosphate deposits were discovered, the phosphate fertilizer industry is facing a big challenge: it must identify and promote efficient fertilizer and crop production techniques that are environmentally friendly and appropriate for each region.

Human society does not want any more to ingest toxicogenic substances, to inhale stale air, or to drink polluted water. No one wants to live, work or relax in environments that are hazardous to human health.

A new form of revolution is being pushed forward, with more far-reaching consequences that the preceding one. To the extent that it is changing drastically the treatment and conservation of our natural resources and the disposal of waste elements necessary for the satisfaction of our vital needs, it is going to modify the organisation of the whole of society in the short and long term.

Until recently, interest in environmental protection was characteristic of a few countries and regions and a main feature of some devoted elite and intellectual associations. However, a mass environmental movement is now progressively shaping the conduct of society and generally but unequally affecting all its components.

It will weigh heavily on the economic equilibria, modify the power structure, and extend or reduce state sovereignty.

For the IMPHOS phosphate industry, environmental policies and the accompanying regulations are both the source of a crisis and the opportunity to get out of it. Originally, the crisis became apparent well before the environmentalist movement began, more precisely in 1970 with the publication of a paper on the toxic nature of cadmium by two Japanese scientists by the name of Yamagata and Shigematsu. At the time, the phosphate market was in full expansion. The Centre for Studies and Research on Phosphate Ores (CERPHOS), which was based in Paris, had been conducting research on behalf of the phosphate industry in North and West Africa. The Centre got interested in the subject of cadmium, and the challenges that it might present one day, and took the first research initiatives to identify the chemical substances in phosphates which might someday be considered undesirable. In fact, when the World Health Organization suggested in 1972 some provisional limits on cadmium dietary intake by man, the newly founded IMPHOS realized the attention being given to the role that phosphate fertilizers play in introducing cadmium into the fool chain. It gave the subject full and particular consideration.

Indeed, the seven phosphate producing companies represented by IMPHOS, namely the "Group Office Cherifien des Phosphates" (OCP) in Morocco, the "Compagnie des Phosphates de Gafsa" (CPG) in Tunisia, the "Entreprise Publique de Fer et de Phosphate" (FERPHOS) in Algeria, the "Jordan Phosphate Mines Corporation" (JPMC), the "Industries Chimiques du Sénégal" (ICS), the "Compagnie Sénégalaise des Phosphates de Taïba" (CSPT) and the "Office Togolais des Phosphates" (OTP), have been carefully regarding the problem of cadmium. They have been playing the role of driving force in the world fertilizer industry, made significant contributions to agronomic research, and taken a responsible attitude to this matter of public concern which stretches beyond their commercial interests: the promotion of an environmentally sustainable agriculture free from toxins.

From phosphate rock mining to finished fertilizer production, the IMPHOS phosphate industry has made great efforts and investments to comply with new, particularly European environmental standards and regulations. Thus, as soon as the first draft of an EEC Directive was elaborated in 1982 to reduce the content of undesirable elements, including the cadmium spread in phosphogypsum, some companies with IMPHOS backing drew a strategy for the elimination of cadmium and the minimisation of some of the attributed undesirable effects. They proceeded in two phases:

- 1982: issuance of a literature review entitled *Dossier cadmium,* which was distributed to professionals in the industry;

- 1984: start-up of a research programme on cadmium removal from phosphate rock and phosphoric acid, funded successively by the World Bank and the Commission of the European Communities (CEC).

Starting from this latter period, the CEC adopted a resolution on 25th January 1988 outlining an action programme for the Community with the aim of reducing the introduction of cadmium in soil from fertilizer by:

- promotion of research/development on various aspects of cadmium in raw materials used for phosphatic fertilizers manufacture;

- development of a strategy to reduce the input of cadmium to the soil, based on an appropriate technology.

The resolution was complemented with more stringent protective measures proposed by Denmark. In response, IMPHOS elaborated a plan of action and adopted the following guidelines:

- more intensive scrutiny of the status of environmental legislation in the EC countries relevant to phosphate rock and downstream products;

- increasing the CEC officials' awareness of the environmental concerns of producers of phosphates, members of IMPHOS;

- implementation of a reasonable research programme responding to the specific environmental needs of Europe.

The cadmium research programme was sponsored by the CEC and started in 1992 with the review of the following reports prepared for the Commission by European experts, upon request from IMPHOS:

- Inventory of studies and processes pertaining to the elimination of cadmium from phosphoric acid, P. Becker - May 1989;

- Review of environmental fate and exposure to cadmium in the European Environment, A. Jensen and F.B. Rasmussen - February 1990;

- Cadmium extraction from phosphate rock concentrate - G. Baudet - May 1989;

- Evaluation of the sources of human environmental contamination by cadmium, ERL - February 1990.

- Inventory of the studies and processes pertaining to the elimination of cadmium from phosphoric acid, A. Davister - July 1992.

- Cadmium removal from phosphate rock (an update), G. Baudet - September 1992.

- Heavy metals in phosphates - Risks to the environment and strategic implications, P.A. Maxson and G.H. Vonkeman - October 1992.

Following a thorough and exhaustive review of an inventory of studies and processes, the CEC expert focused the attention of CEC and IMPHOS on a process developed by CERPHOS for cadmium removal from phosphoric acid. The co-crystallisation of cadmium with the anhydrite process has been subject to further research, with the Commission allocating funds for the project.

From mid 1994, the first bench-scale of the project, which has just been completed, served to demonstrate the technical feasibility of the CERPHOS process. The regular and rigorous monitoring of this phase by the CEC expert, and the encouraging advice he offered to the CERPHOS-IMPHOS team, testify to the interest manifested in this project and gives IMPHOS Member Companies reasons to hope that phase II of the project, involving pilot-scale research, would benefit from the same moral and financial support of the European Union.

Indeed, having successfully achieved the objectives of phase I of the project, additional information will be needed to design plants able to eliminate cadmium from all types of phosphoric acid and to evaluate their cost, as stipulated in the IMPHOS-EU Agreement.

Therefore, it is essential to use the data from the first phase to design a semi-industrial pilot plant. This is phase II of the project, which involves continuous operation of the process in order to be able to withdraw all the information needed. That way, an adequate and definitive solution to the problem of cadmium in fertilizer would be made available.

Considering the key role played by the European Union in the area of environmental protection, we are confident that it will take the right decision to help the project come to an end

For more than a quarter of century, the IMPHOS phosphate industry has been expending much scientific, human, financial and material effort trying to address the environmental issues that relate to phosphate rock mining and fertilizer manufacture. Beyond this effort, which will undoubtedly prove worthwhile, IMPHOS plans very soon to create a body to "watch over technological and environmental development" in the manufacturing of fertilizer. While it is hard to predict the outcome of the environmentally driven legislation, it is indispensable to make use of this "watch body" to assess and predict technological and environmental impact on the fertilizer industry and world farming.

To this effect, IMPHOS will endeavour to establish contacts and conclude agreements with all national, international and regional organizations with a common interest. Furthermore, the Institute is considering setting up an experimental network to determine accurately the cadmium balances for a range of soils and cropping systems in different parts of the world (Europe, Asia and Latin America, in particular) where intensive farming is practised. With high phosphate applications, there must be an effort to monitor the cadmium level in soil and plants and determine the effectiveness of government, industry, and farming practices in decreasing the level of cadmium in plant-based foodstuffs. Experimentation should permit the construction of cadmium balances for the existing crop production systems and the collection of reliable data that will help advance our knowledge of the cadmium cycle. This scientific information can then serve as a basis for taking appropriate agricultural and regulatory decisions.

At this turning in the history of agriculture, and while it is easy to see the rationale behind the application of new environmental standards (Table 1), IMPHOS cannot help worrying about the negative impact, if such standards become too restrictive, on the profitability and trade volume of its Member Companies.

It should be noted that there is insufficient low-cadmium rock phosphate available in the world to meet the levels of rock required to ensure adequate food production. Should, for instance, a strict limit on cadmium content in fertilizer be imposed in the OECD countries, as advocated by some (Table 2), it would undoubtedly have devastating economic and environmental effects. It would be impossible to use rocks from traditional OECD suppliers. Six IMPHOS members, in addition to companies in the USA and Israel, would be excluded from the OECD market with:

- sequential shut-down of plants and mines;

- massive layoff of the work force;

- soil phosphorus depletion and loss of fertility;

- setback of agricultural research and development;

- ultimately, soil reserves of phosphorus reaching minimum levels by the year 2005, which would jeopardize the global food supply.

Fortunately, this pessimistic scenario is not likely to happen any time soon. No one among environmentalists, much less in the sphere of phosphate producers and farmers, would believe in such a scheme. It is true that cadmium is toxic, but it remains equally true that no scientific evidence exists to substantiate the claim that cadmium in fertilizer is toxic to plants or the animals and humans who feed on those plants.

Existing data indicate that if there is a problem, it is of a long-term nature. Bockman et al. (1990) calculated that if phosphatic fertilizer was made from a phosphate rock with 25 mg cadmium/kg and was then applied to supply 20 kg phosphorous per hectare (46 kg P_2O_5/ha) each year, it would take 100 years for soil cadmium in the topsoil to increase by 0.4 mg/kg. In other words, it would take seven centuries for soil cadmium to increase by 1 mg cadmium/kg, which is far from justifying any regulation of cadmium in fertilizer.

At the request of the EC Commission, the European Fertilizer Manufacturer's Association (EFMA) summed up the situation in 1989 as follows:

"For the whole European Economic Community, the average cadmium content of fertilizers is 60 mg/kg P_2O_5, which means that if the distribution of fertilizers was uniform, the amount of cadmium applied would be 2.5 g/ha/year or less than 1 microgram cadmium per kg of cultivated soil. Knowing that the average cadmium content of soils is 0.5 mg per kg, 500 years would be needed to double the soil content if no cadmium was taken up from the soil."

Finally, from the long-term Rothamsted experiments, A.E. Johnston and K.C. Jones (1992) found out that the amount of cadmium applied in the annual dressings of 400 kg/ha of simple superphosphate has averaged 2 g/ha/year. This cadmium caused no increase in the cadmium content of either topsoils or subsoils.

However, cadmium had accumulated in acid grassland soils that contained a high level of organic matter and received superphosphate. Liming was an effective management practice for decreasing the level of cadmium in acid soil.

Table 1

Elimination of phosphate sources (suppliers) to the European Union should cadmium exceed

250	200	150	100	50
(mg Cd/kg P_2O_5)				

None	Senegal	Senegal
		Togo
			Senegal
			Togo
			North Carolina (USA)
			Morocco:
			  Boucraa
			  Youssoufia
			Tunisia (P)
			Israel (P)
				Senegal
				Togo
				North Carolina (USA)
				Morocco (T)
				
				Tunisia (T)
				Israel (P)

P: Partially
T: Totally

Source: A.J. Williams

Table 2

Maximum Cadmium Concentration in Phosphate Fertilizers

Country	Limit (mg cadmium/kg phosphorous)
Austria	275 mg
Denmark	150 mg (from July 1992) 110 mg (from July 1995)
Finland	50 mg
Germany	200 mg (voluntary)
Japan	340 mg
Netherlands	40 mg
Norway	100 mg (from 1993)
Sweden	100 mg (there is a fee for concentrations between 50 and 100 mg)
Switzerland	50 mg (from 1992)

In concluding this presentation, I would say that the regulation of maximum cadmium concentrations in phosphatic fertilizers which has been under consideration in various countries is the result of environmentalist dynamics which reel off admittedly serious presumptions, yet they do not measure up to indisputable facts. In this regard, it is important that the regulation that some are trying to impose is sound and scientifically based.

For our part, we are convinced that in order for these dynamics to proceed in harmonious, balanced and rational patterns, they must be in line with the findings from scientific research and agronomic experimentation. Established facts and constantly renewed dialogues, such as the one taking place in this workshop, would hopefully lead to a consensus and a unanimous compliance with the final resolutions.

Again, IMPHOS members are keen to take a responsible attitude to matters of public concern. They do their best to meet the European and global environmental standards.

They have been making great progress in this regard and will continue their efforts, but it is important that the phosphate industry not be burdened with unnecessary new cost, considering the low profitability of this industry. Furthermore, any new standards and any legal instruments to enforce them should only be considered as and when a viable and effective cadmium removal process is commercially available at a bearable cost.

REPORT OF SESSION B

IMPLICATIONS OF MEASURES TO REDUCE THE LEVELS OF CADMIUM IN FERTILIZERS

The papers presented in this section were given during Session B.

Introduction

The task was to examine the implications of measures and techniques to reduce the cadmium content of fertilizers within OECD countries and phosphate exporting countries, including benefits, costs, trade effects, economic/socio-economic impacts, need for technology transfer, investments, etc. Discussion started by examining the range of potential issues that could arise from measures to lower cadmium in fertilizers. The complexity of these issues required limiting the scope of the discussion, and it was decided to develop a plausible hypothesis of reducing the cadmium content of fertilizers so that the relevant implications could be examined.

Taking into account that the EU is, as a consequence of the accession agreements with Austria, Finland and Sweden, in the process of reviewing the issue of cadmium limits in fertilizers, it appeared that this might be the most helpful route to follow in our discussions. There was no discussion on the appropriateness of the measures or the science used to justify such measures. The example simply sets some bounds on the discussion. For this reason it was decided that a cadmium content limit of 60 mg cadmium/kg P_2O_5, effective from 1 January 2000, should be chosen as the starting point for the case study. It was agreed to include in the discussion views of how the impact of these measures would vary if the limit was phased in rather than implemented immediately. Similarly the group agreed to reflect, where appropriate, on the effect of variations in means of implementation. It was stressed that the use of this example was neither an agreement nor a prediction of the approach or level, but simply a convenient starting point for discussion.

Possible modalities for the decision

A number of direct measures were identified:

- legislation (e.g. an EC Directive);

- voluntary agreements or commitments (which can achieve similar effects as legislation, if they are comprehensive); and

- labelling requirements applicable to both of the above.

There are indirect measures that can be used:

- Economic instruments are an option that can be used both alone or in combination with technical requirements;

- Public pressure as a political reality; and

- Education on appropriate use of fertilizers containing cadmium.

Implementation

Implementation could be carried out by national governments that create legislative measures or by the private sector, which could enter into voluntary agreements.

Impacts

It was recognized that numerous more or less probable impacts could occur from the hypothetical action of lowering cadmium levels in fertilizers in the EU. These were discussed under the following several categories.

Access and Equity

Producer countries can suffer effects based upon the concentration of cadmium in rock. Producers of high cadmium rock will lose markets and those with low cadmium rock could gain markets. There is a potential for increased depletion of low cadmium rock deposits that could lead to potential supply shortages in the short term.

In producer countries there may be a shift in the structure of the industry from traditional fertilizer production towards bulk blending facilities that utilize other fertilizer intermediates.

In European Union countries taking the measures outlined in the working hypothesis, price increases may occur for phosphatic fertilizers. This could raise consumer costs for produce in some of the EU countries.

In other countries, segmentation of fertilizer markets could occur through the development of a price differential for low and high cadmium fertilizers. A change in market flows would then occur, forcing a greater use of high cadmium fertilizers in non-EU countries. Product differentiation may also occur. Public pressure could be placed on the agricultural sectors of countries utilizing high cadmium fertilizers, thereby discriminating against their products. This would, in turn, cause displacement effects in these countries. A number of participants thought the outcome of these changes could be higher food production costs and higher consumer costs or lower agricultural production rates.

Costs of Production

No general economically viable process has yet been found to remove cadmium from rock. Thus cadmium removal is likely to occur only during the fertilizer manufacturing process. It was noted that various methods of removing cadmium in phosphoric acid are currently being examined. While these processes show promise, it is estimated that 7 to 10 years will be required to bring these facilities on line. This process would affect about 70% of the world's P_2O_5 consumption. Other parts of the fertilizer industry, not using wet process phosphoric acid, do not currently have a technological alternative other than the use of low cadmium rock, or blending to reduce overall concentrations. Achieving the desired concentration by blending various rock types might involve the addition of blending equipment to facilities that are not currently equipped with such facilities, and this will result in increased costs.

A single design/technology for removal of cadmium is unlikely to be effective in all cases. The influence of different rock types on the chemical process may limit the utilization of any single removal process.

There is likely to be an increase in production costs of $15-$30/tonne P_2O_5 or according to the consultant's report, a farm gate increase of 1-7 per cent. Restrictions or bans on the use of high cadmium fertilizers could improve the economy of cadmium removal technology.

Environment

The most important environmental effect of the implementation of the EU measure could be a drop in the flow of cadmium to agricultural soils in Europe.

Numerous environmental effects specific to individual countries could be identified. For instance, in producing countries measures to remove cadmium will increase the generation of "Basel" wastes. If these countries do not handle these materials in an appropriate manner, serious environmental and health effects could occur.

The displacement effects mentioned above can lead to the use of higher cadmium fertilizers in countries that are not able or willing to impose the same restrictions, thereby increasing the flow of cadmium to agricultural soils.

Trade

Under the working scenario, low cadmium phosphatic fertilizer could attract a price premium or an increased market share. Prices of high cadmium fertilizers could be expected to fall. The price differentiation would result in changed trade flows. Countries and companies with access to low cadmium deposits would benefit, while others would lose. Senegal, Togo, Morocco, Tunisia, Nauru, Israel and North America would be likely to be adversely affected.

The extent of change will depend upon the level of the cadmium limit, and on whether blending is feasible and commercially practicable. Ultimately trade would be skewed in favour of those countries with access to, and the ability to apply, new technologies for removing cadmium from raw materials and intermediates.

It is unlikely that such restrictions would give rise to international trade rule concerns, or to obligations under the Treaty of Rome, provided the measure can be justified in terms of legitimate environmental or human health policy objectives.

There is a concern that the implementation of a restriction on the use of high cadmium fertilizer could lead to consumer preferences for imported food based upon perception rather than the cadmium content in the food.

Competition

There would be similar effects on companies. Those with access to low cadmium deposits would benefit, others will lose. Competition will be built for the low cadmium rock, thus promoting a change in the pricing structure. Both these factors could lead to product and price differentiation and market segmentation that could affect investment flows and opportunity costs.

Consequential Effects

It was recognized that there was a potential for consequential effects of an EU measure, in particular the possibility of other countries implementing similar measures. This could exaggerate some of the effects noted above, but the group did not address any of these effects in detail.

Agronomy

Some participants thought it was important to note that there are potential agronomic effects. For example, if the effect of EU action were to be followed by wider implementation, and rises in the cost of phosphorus fertilizer due to technical costs of cadmium reduction, fertilizer consumption could decrease and impacts could be seen in the use of nitrogen and potassium fertilizers. This could have effects in other countries that would cause:

- unbalanced application of plant nutrients;

- reduced nutrient use efficiency (especially nitrogen); and

- depletion of plant nutrients from soil reserves, resulting in losses in crop yield and quality, and effect on environmental protection.

Mitigation

Design of Measures

The implications of measures are highly dependent upon the way the requirements are designed. Some examples of different mechanisms are:

- presenting requirements as minimum or average figures;

- using economic instruments or requiring strict limit values;

- acknowledging regional differences within the EU; or

- utilizing different labelling requirements.

It was noted that a clear definition of relevant analytical methods would be useful in any case.

Blending

One way of mitigating many of the effects in the short term is to blend the rock, or other fertilizer feed stocks, to achieve the desired cadmium level.

Technology Development and Transfer

There are some obligations for developed countries to provide technical assistance to developing countries. To meet the expected increase in demand for low cadmium in fertilizer, the EU has supported and given assistance to producing developing countries to develop the cadmium removal technology discussed above. This will increase the relative competitiveness of the companies participating in the development programme. One consequence of such activity might be that other companies in various countries will suffer if the relevant technology is not made available to them.

Timing of the Implementation Measures

Clearly, if the implementation of the measures were to be co-ordinated with the commercialization of new technology, impacts would be reduced. In particular, it would allow countries with high cadmium rock time to adjust. Not putting short-term pressure on the availability of low cadmium fertilizer would limit the price differential and help minimize distortions in market flows, or economic and environmental impacts on non-EU countries.

Making the technology available by providing appropriate licensing agreements before the measures are put into place would contribute to mitigation of major displacements. Careful thought should be given to the timing of the solutions in relation to the time frame for addressing the problem.

Agricultural Practices

The effects of the EU measures on increased use of high cadmium phosphorus fertilizers in other countries, as a result, could be mitigated by adoption of agricultural management practices which may reduce cadmium bioavailability.

Adjusting application rates in countries that have high soil phosphorus levels may offset fertilizer price increases as long as the N:P:K ratios come into line with acceptable levels.

Conclusions and Recommendations

1. Potential benefits of such a measure would include technical innovation and reducing the input of cadmium to agricultural soils in Europe.

2. However, measures to reduce the use of high cadmium fertilizer would raise many complex issues.

3. The implications of such measures could be significant, and manifold.

4. Many of these implications would go beyond the countries immediately involved.

5. There is a range of strategies for reducing adverse impacts, which cannot all be avoided.

6. If such measures are to be introduced, careful thought should be given to both the costs and benefits of these measures, and to the timing of the solutions in relation to the time frame for addressing the problem.

SESSION C

ACCUMULATION IN AGRICULTURAL SOILS AND CONTENT IN FOOD AND HUMAN UPTAKE

Current Developments in the Use of Fertilizer Phosphorus and the Consequences Concerning Cadmium

Göte Bertilsson

Hydro Agri AB
Landskrona, Sweden

Abstract

Cadmium input via fertilizers depends both on the amount of fertilizer applied and the cadmium content. The application of phosphorus has declined in most industrialized countries during the last decade. The change in trend has two main reasons:

1. During the main part of this century, agricultural soils have been enriched with phosphorus by applications widely exceeding the export by crop products. This *enrichment phase* has improved the phosphorous status of soils, and the yields can be sustained by lower inputs. Gradually, agriculture can change to a *replacement phase*.

2. There has been improved utilization and valuation of manure phosphorus.

A replacement policy cannot be applied rigorously in each individual case. There are varying soils and crops to consider. Local research and experience should always be the basis. The development, however, seems clear: industrialized countries with positive soil phosphorous balances will continue to reduce phosphorous inputs and go towards a replacement. On the other hand, several developing countries have a negative soil phosphorus balance in combination with poor soils. There, the fertilizer phosphorus use will increase. This change in pattern will have implications for the cadmium additions to soils. In both cases it is important, along with continued development concerning recycling, efficient fertilizers and application methods, and balanced agricultural systems.

The example of Sweden

The cadmium input by fertilizers depends on both the cadmium content and the amount. Both have been changed in Sweden during the latest decades. Both factors are also important for the work on system improvement.

The effects of different combinations of cadmium content and phosphorus intensity are shown in Table 1.

Table 1

Addition of cadmium in grams per hectare and year at different combinations of phosphorus application and cadmium content of the fertilizer

	120 mg Cd/kg P	30 mg Cd/kg P
P as in 1970 (25kg P/ha)	3	0.75
P as in 1995 (8 kg P/ha)	0.96	0.24

Phosphorous use in Sweden

Phosphorous use in Swedish agriculture is shown in Figure 1. It is expressed in kg fertilizer phosphorous per hectare of agricultural area. Around 15 per cent of this area is currently set aside. Consequently, for the latest years the intensity on active land is 15 per cent higher than shown in the diagram. This is of little importance for the discussion here.

Figure 1

Consumption of phosphorous in Sweden

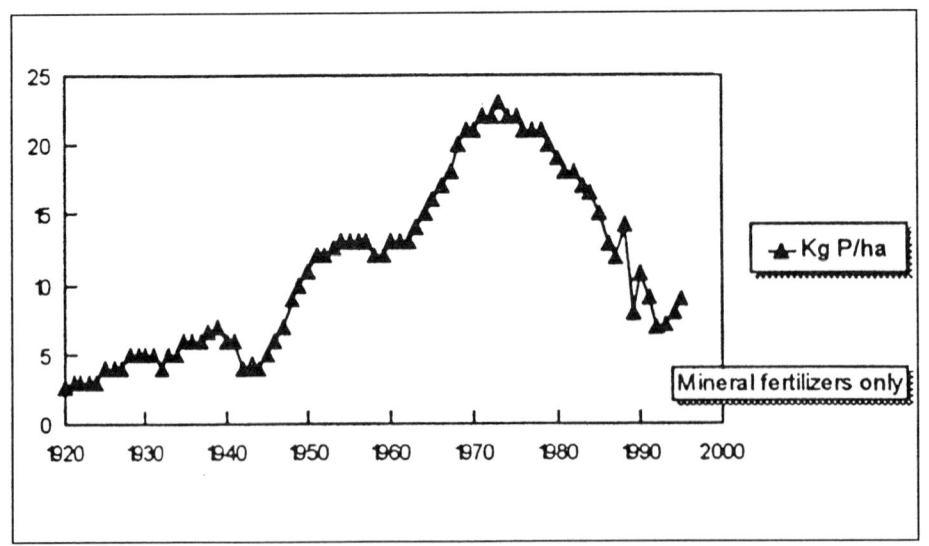

Phases of phosphorus use

Up to 1970, phosphorous use steadily increased. At the same time the content in the soils also increased as measured by the method normally used in Sweden (P-AL). We had an *enrichment phase*. The additions of phosphorous to the soils exceeded the export by crops.

For the development of Swedish agriculture, this was a necessary phase. In the first half of the century the soils were poor in phosphorous and additions brought about dramatic yield increases. This also led to the importance of phosphorous becoming deeply rooted in the agricultural society.

The enrichment phase was important, but it lasted too long.

If we look at the latest period, roughly since 1985, the fertilizer phosphorous is 7-8 kg per hectare. Together with manure phosphorous (on average, 6 kg per hectare), the additions are about as large as the export by crop products. We have a *replacement phase*, where the additions are about one third of those 20 years ago.

In most industrialized countries the use of phosphorous has started to decline, and there is a transition from an enrichment phase to a replacement phase. In Sweden this development started earlier and is more evident than in most other countries.

Factors behind the change

The physical background is that the phosphorus supplying capacity of the soils has increased because of the enrichment, and consequently less fertilizer phosphorous is needed. The recommendations tend towards replacement at high soil phosphorous levels. This will be discussed further later on.

If we compare Sweden and surrounding countries, some differences are worth pointing out:

- Swedish advisors and farmers are especially sensitive to production economics and marginal economic return. The sharp rise in phosphorous prices in 1974 lead to a re-thinking about phosphorous application.

- The environmental discussion started early in Sweden. In combination with the higher prices, this enforced a re-thinking. The taxes on phosphorus imposed in 1988 made oversupply very costly.

- Much of the "oversupply" was caused by the neglect of manure as a phosphorous supplier. Manure was often seen as a "premium" in addition to normal fertilizer supply. The structure of animal production made development of correct manure use easier than in surrounding countries (density not too high). Consequently, more fertilizer phosphorous could be replaced by manure.

- A consensus on fertilizer advice enhanced the development.

The phosphorous in soils: recommendations and the replacement principle

In order to gain a perspective on possible development in the future and in other areas, a short summary of phosphorous in soils would be advantageous.

Figure 2 shows the soil phosphorous fractions in a soil with different fertilizer treatments for 27 years (soil fertility experiment, Orup). Three fractions are shown: 1) "available" phosphorous, extracted according to the AL method; 2) less soluble inorganic phosphorous, extracted by 1 N HCL; 3) organic phosphorous (total phosphorous minus fractions 1 and 2).

Figure 2

Consumption of phosphorous in Sweden

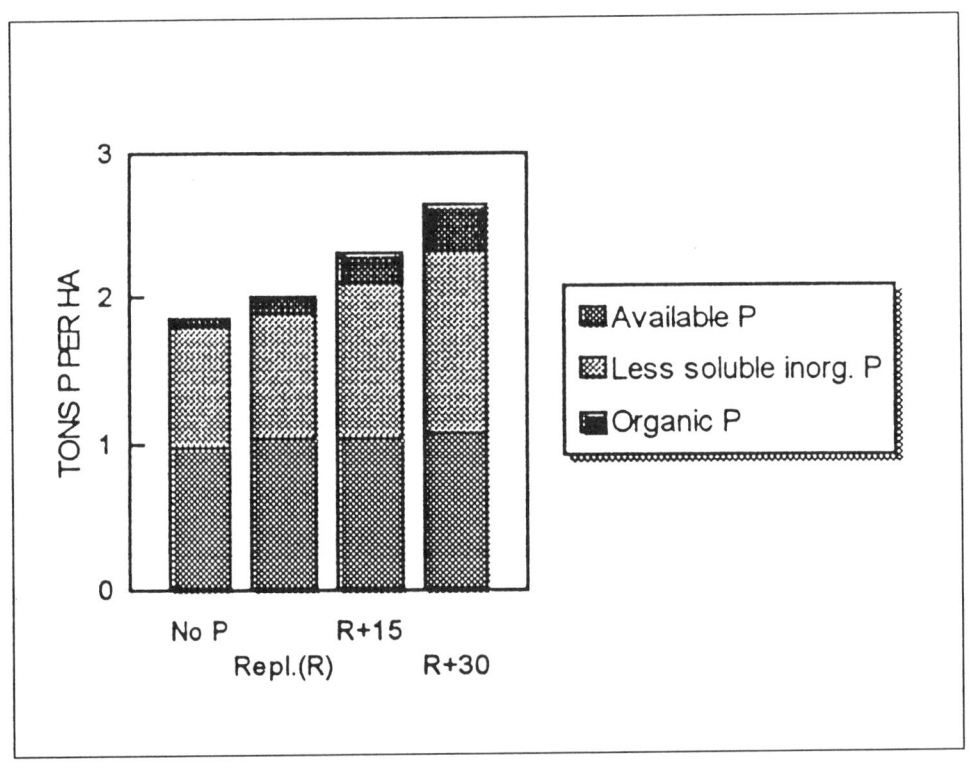

The relations between the fractions ilustrate that the added fertilizer has to some extent increased the less soluble inorganic phosphorous, and also that long-term crop growing without fertilizer has diminished the less soluble fraction. There is a kind of equilibrium between the different fractions.

Arithmetic replacement of the export by crop products will not suffice to keep up the available phosphorous if some of it is "fixed" in the less soluble fraction. On the other hand, if the soil pools are high or active enough, the available phosphorous can be kept for some time without additions. Much agricultural phosphorous research is engaged in quantifying these relations for different soils and crops and improving the background for fertilizer recommendations.

The current phosphorous recommendations issued by Hydro Agri Sweden are given in Table 2.

Table 2

Phosphorous recommendations Sweden (Hydro Agri)

Cereals	Recommendation kg P/ha in different soil P classes				
Yields t/ha	I (very low)	II (low)	III	IV	V
2-4	30	20	15	10	0
4-6	35	25	20	15	0
6-8		30	25	20	5
8-10			30	25	10

The recommendations include manure and other recycling of phosphorous.

The recommendations in class III are a little higher than replacement phosphorous content in harvested grain, and those in class IV are a little lower. At higher phosphorous levels in soils the recommendations mean replacement of utilization of accumulated reserves.

However, this is not the case for all crops. For potatoes in class IV, 50-60 kg phosphorous is recommended although the harvested potato does not contain more than 20 kg. The potato is a poor phosphorous feeder and needs high levels of available phosphorous for optimal performance.

The crops differ in their requirement of available phosphorous. The base is root morphology and root physiology. Also mycorrhiza can play a role and enhance the phosphorous uptake.

The replacement principle cannot be used for individual crops, but as a balance over a rotation.

Soils phosphorous balance

The important components in the balance are the following:

+ fertilizer phosphorous

+ recycled phosphorous (manure most important, in addition to sludges, etc.)

- phosphorous taken away in harvest products

- leaching and erosion (quantitatively unimportant for the soil balance in European conditions).

Both a continuous positive balance and a negative balance are unsustainable in the long run.

The positive balance can be adjusted by reduced fertilizer applications unless the load of animal manure, etc. is too high. It should not be neglected that some soils in Europe still need enrichment and a positive balance. The time for a replacement phase may not yet have come.

A negative balance may seem acceptable for a time: for instance, for efficient crops needing no or little fertilizer, or "ecological" farming systems utilizing previous build-up. However, the only measure which can give a sustainable situation is replacement. Within the total amount of phosphorous applied, the fertilizer can be reduced if recycling can be increased. This is the way to go, and recycling development shoud lead the way.

Techniques for improved phosphorous utilization

Techniques for improved phosphorous utilization may reduce the need for enrichment. A theoretical example: cereals, 6 tonnes, class II need 30 kg phosphorous according to Table 2. With phosphorous replacement, maybe 20 kg suffices, which means that the replacement phase could start in class II.

Work is going on with:

- replacement as such;

- replacement of multinutrient fertilizers with combination effects;

- fertilizers with improved efficiency adapted to the situation (e.g. banded polyphosphates);

- foliar fertilizing of some crops (potatoes).

Rotations, agricultural systems should be adapted to utilize residual phosphorous from "high phosphorous crops" like potato. If potatoes are grown too often, a continued enrichment will result. This is not an easy discussion in regions with high density demanding special crops. Improved fertilizer techniques might help.

It is important that the agricultural extension and advice given can establish confidence in relation to the farmer. A fertilizer use adjustment is for his long-term benefit. The concept of "replacement" cannot be applied rigorously. There are varying soils and varying crops to consider. Local research and experience should always be the basis.

Future phosphorous use development

In Europe there is a development towards reduced phosphorous application. This will undoubtedly continue, since the rates still considerably exceed replacement. The locally large excess of phosphorous in manure is an obstacle in the development.

In Africa, for instance, there is the opposite situation. Nutrient depletion is a key problem. Both recycling and fertilizer use need to be increased. There an enrichment phase needs to get started. A burning problem is whether this can be afforded in relation both to local agricultural economy and the world phosphorous reserves.

Some cadmium/phosphorous relations

In relation to fertilizers, the cadmium content is often expressed as mg Cd/kg P. This is a practical unit for comparing different materials.

Just for comparison and food for thought, some more Cd/P relations:

Swedish soils. Average value: 550 g Cd/ha. Amount of phosphorous in topsoil about 2000 kg. These figures give 275 mg Cd/kg P.

Wheat. According to market demands in Sweden, the cadmium content shall not exceed 100 micrograms/kg. A normal phosphorous content is 0.3 per cent. These figures give a value of 33 mg Cd/kg P.

Conclusions

The cadmium input via fertilizers depends on both the amount applied and the cadmium content.

Previously high phosphorous applications in industrialized countries are being reduced. A necessary enrichment phase is slowly developing into a replacement phase. The development might be long-term and slow, with local modifications.

On the other hand, enrichment needs to be started in several parts of the world.

This change in pattern will have implications for the cadmium additions to soils.

In both cases it is important, along with continued development concerning:

- recycling;
- efficient fertilizers and application methods;
- balanced agricultural systems.

References

Gunnarson, O. 1987. Den långsiktiga fosfordynamiken i de skånska bördighetsförsöken. Kungl. Skogs-o. Lantbruksak. tidskr. suppl 19.

Jansson, S.L. 1975. Long-term soil fertility studies (Swed.). Kungl. Skogs- och Lantbrukssakad. Tiäskr., Suppl. 10.

Junck, A., Claassen, N., Schultz, V. and Wendt, J. 1993. Pflanzenverfugbarkeit der Phosphatvorräte ackerbaulich genutzter Böden. Z. Pflanzenernähr. Bodenkunde 156, 397-406.

McCollum, R.E. 1991. Buildup and decline in soil phosphorous: 30-year trends on a typical Umprabuult. Agron. J. 83, 77-85.

Nielsen, J.D. 1994. Crop recovery of fertilizer phosphorous from soils low in soluble phosphorous. Acta Agric. Scandin. 44, 2, 84-88.

Cadmium in Mineral Fertilizers

Johannes Dettwiler

Office for Environment, Forest and Landscape (OEFL)
Switzerland

1. The Federal Institute of Environmental Protection and Agriculture (IUL), Liebefeld-Berne, Switzerland, has for 1989 carried out a comprehensive survey on the cadmium content of fertilizers. In a joint report of November 1991, the IUL and the OEFL have published the results.

2. In the said year, 1 tonne of cadmium caused by mineral fertilizers was set free on Swiss soils (estimated intensive crop and pasture area: 600,000 ha). This compares to 0.5 tonne of cadmium caused by sewage sludge application, and 0.7 tonne of cadmium in the total mass of animal slurries (according to a survey in 1994-95).

3. The Ordinance of June 1986 on hazardous substances (OHS) carries relevant statutory measures to ensure fertilizers being applied with care and moderately (articles 9 and 10, OHS).

4. Anyone who uses fertilizers must take account of the nutrients in the soil, of the nutrient requirements of the plants according to general guidelines for fertilizer use, of the site conditions where fertilizers are used, of the weather conditions, and of the restrictions imposed locally by authorities to protect waters or nature conservation areas (appendix 4.5, OHS).

5. Fertilizers mentioned below may only be supplied to users, if:

   - mineral fertilizers are conforming to a maxium cadmium level of 50 g/t of phosphorus (as from January 1, 1993; notified in 1991 at EFTA in Brussels, 91/9058/CH);

   - sewage sludge is conforming to a maximum cadmium content of 5 g/t of dry matter (as from October 1, 1992);

   - compost is conforming to a maximum cadmium level of 1 g/t of dry matter (as from September 30, 1995).

6. Furthermore, within any period of three years, not more than 25 tonnes of compost or 5 tonnes of sewage sludge dry matter shall be applied per hectare, provided the nitrogen and phosphorus content in such fertilizers shall allow for that quantity at all.

7. The Ordinance of June 1986 on pollutants in soils (OPS) sets guide values for heavy metals in soils. With respect to cadmium, the soil is considered fertile when conforming to a cadmium content of less than 0.8 g/t of soil (HNO_3 extract) or else to a soluble cadmium content of less than 0.03 g/t ($NaNO_3$ extract).

The Role of Farmers' Organizations in Reducing Cadmium in Food

Jan Eksvärd

SLR-Swedish Farmers Supply and Crop Marketing
Organisation, Svenska Lantmännen
Stockholm, Sweden

The farmer's role in their organization

Swedish farmers organized themselves in co-operatives for purchasing and marketing about 100 years ago. Collecting and marketing of grain, milk, etc. was the first objective, followed by collective purchasing of needs like fertilizers, seed, tools, machinery. Today approximately 80 per cent of the farmers are members of co-operatives. Normally, each of them is a member of five cooperatives. The co-ops are grouped into 16 different branches. Examples of such branches are dairies, banks, distillers, oilseed producers, and *Lantmännen* (supply and crop marketing). The market shares of the co-operatives in their respective business areas are mostly between 60 and 95 per cent.

Lantmänn sells all equipment, feed, fertilizers, etc. the farmer needs for his production of vegetable or animal produce, and collects cereals and oilseed. *Lantmänn* also owns companies that process the cereals and oilseed, such as the Cerealia group that mills over 50 per cent of the cereals in Sweden and Denmark and is a leading supplier of bread.

Farmers rule their co-operatives and companies by boards and hold yearly general assemblies. The objectives for the co-operatives are to give the farmer high prices on produce and low ones on supplies. Today these objectives are also linked to future economic sustainability. For example, high contents of cadmium in phosphates can increase the level in the soil, and in wheat for bread, and reduce quality to an unacceptable level. Thus, the fields may become impossible to use for food production. One of few examples, yet, is batches of flax seeds from west Sweden which could not be used for food.

Motives to reduce the cadmium content

In 1988, SLR became aware of proposals for lower limits for cadmium in wheat. Later 0.1 mg/kg wheat was suggested by WHO, the EU and the Nordic countries. While analysing the situation, we realised that we only knew medium values from a few samples in some areas. These values gave very small safety limits for the suggested limit value, compared to the margins for pesticides or other metals. Possible variation between areas, varieties and growing techniques was not known. What advice could be given to farmers with high levels of cadmium? Which forms of cadmium in different food were dangerous and penetrated into the human body? What could we do to reduce the addition of cadmium to the soil? Should we act at all? Could the farmers not trust the authorities and wait until guidelines were given and all limits set?

The experiences from the pesticide debate show that consumers do not fully trust authorities. As consumers we take our own stands from fragmentary information, and from those views and feelings we might have on that information. For each and everyone this is very relevant and true. We then act on our stand when we buy food or talk to friends, colleagues, etc.

Obviously, farmers could not wait if they were interested to maintain high confidence in their products. So a number of reasons motivated the farmer's organisations to act:

1. Lack of knowledge on whether all regions could meet the new limit values for cadmium in wheat or not. How did the contents of cadmium in different crops, varieties and regions vary? What products needed extra low contents of cadmium?

2. No plans for actions existed, and a base for such actions was needed.

3. The farmers had contributed, and were still contributing, to cadmium accumulation through phosphate fertilizers, feed phosphates and sewage sludge.

4. They wanted market advantages for Swedish food compared to imports.

Actions taken and results

At first, interest was focused on mapping the variations in cadmium contents of cereals and lowering the concentration in phosphorus fertilizer.

The mapping started in the south and west of Sweden. Correlation between concentrations in the roots of watermoss and in wheat was shown, but only as mean values over large areas. In fact, in a high content area samples with the lowest content in wheat could be found. The opposite was also true. It was necessary to continue the mapping and follow each field for a number of years. Skånska Lantmännen in south Sweden has analysed samples of all wheat transports from farms for three years. From this background, information will be available concerning effects of growing and fertilizing techniques, content in varieties, and yearly fluctuations.

When a farm is found to be in the risk zone for producing crops with high cadmium levels, a personal visit is necessary. The farmer and representatives from Lantmännen can analyse the situation and discuss possible alternatives. Of course this is private, since the value of the land is at stake.

The contents in fertilizers could vary from almost zero to over 250 mg Cd/kg P. A normal dose of 25 kg P gives 6.25 g Cd/ha. The normal loss is approximately 0.7 g Cd/ha. This accumulation was not acceptable, and in the beginning of 1990 the central purchasers in SLR decided not to buy any fertilizer with cadmium content over 100 mg Cd/kg P. This resulted, of course, in lowered mean values. Today the average content of cadmium is 30-35 mg Cd/kg P. The Swedish tax on cadmium in fertilizers of over 5 mg/kg P supports the ambition to keep the content at this level.

The Federation of Swedish Farmers, LRF, stopped the spreading of sewage sludge on cultivated land in 1990. The reason was high contents of metals and uncertainties about anthropogenic chemicals in the sludge. Now, after almost five years, an agreement has been reached between Swedish EPA, LRF and VAV (Swedish Water and Wastewater Work Association) about metals and organic chemicals in sewage sludge. This means that the cadmium load in the year 2000 is maximized to 0.75 g Cd/ha and year, compared to the suggested level of 2 g Cd/ha and year. This is a breakthrough in efforts to reach a balance between addition and reduction of cadmium to cropland. Yet another 50 per cent reduction will be necessary to reach the goal balance.

Differences in cadmium uptake between varieties is a way both in short- and long-term perspective to reduce the contents in the food. Since 1990 a number of trials comparing varieties have been performed. Today varieties with high contents of cadmium are avoided in areas with high content of cadmium in the soil. In the development of new varieties, cadmium uptake is one new parameter for selection. In food crops, low-collecting types are preferred. In Salix, a populous variety grown for energy, high-collecting types could be of interest. During the 25-year growing period, up to 50 per cent of the cadmium content in the soil could be removed. Of course the metal is found in the ashes, which need separate treatment.

A high content of cadmium in a certain type of food is not correlated to high uptake in the body of warm-blooded animals. Projects to find correlation between type of fibres in food and cadmium uptake are financed from Cerealia Research Funds. In the area of bioavailability more research is needed.

What does the farmer feel about cadmium?

No investigations have been carried out in this area. However, in their hearts farmers generally are eager to keep the land productive and vital for the next generation of farmers.

A small percentage of the farmers have a too high cadmium content in their soils. The farmer is concerned if he knows about it. One way he gets information about accepted levels is through contacts with his customer (Lantmännen). When the levels are measured and are too high, it is hard facts. The farmer may need to change his choice of crops. This could be life or death for him as a farmer. Farmers act like other people in difficult situations.

Concerning environmental fees or taxes on cadmium in fertilizers, the farmer does not want fees that give him a worse economic output. Society (we) has (have) put agriculture in the same situation as other sectors in our part of the world. Agriculture needs fertilizers for high productivity and wants "society" (the fertilizer industry) to provide them at reasonable prices and good quality. To put taxes on cadmium in Sweden is just a cost increase for the farmer. It gives a reduction in cadmium loads only if there are low-cadmium fertilizers available. This is an international question. Collective pressure to clean the fertilizers is one way: farmers' collective actions against, or with, fertilizer producers to increase the request for and use of low-cadmium fertilizers. The prices will increase and reach a level where it will be profitable for producers to clean the phosphate. In this example, some farmers or some countries will for a long time get cadmium-rich fertilizers. From the ethical point of view, this is hardly an acceptable policy.

An alternative is to set the same limit for all western countries and give the industry some years to reach the goals. The residue of cadmium needs to be taken care of in a proper way.

Future needs

Besides accumulation in soils, cadmium is also accumulated in our technosphere (bridges, cars, batteries, paints, etc.) In Sweden 5000 tonnes had been built into structures and products by 1990. This quantity will slowly move through water, air and food to sediments in rivers, lakes and seas. In Swedish agricultural soils the total content today is 1700 tonnes. The 5000 tonnes must not be allowed to reach the food chain. Probably the fluxes of waste water and sewage sludge have a key role in this transport of metals.

More research is also necessary on bioavailability and breeding to prevent damage from already high levels in some areas and for risk groups in the population.

To maintain and, in some soils, to increase the pH is urgent.

Since cadmium is accumulated in the soil, additions from fertilizers, sludge and air need to be reduced from, respectively, 0.7 g/ha (based on total phosphorus to 50 per cent of total acreage) + 1 g/ha (3 per cent of total acreage) + 0.5-1.5 g/ha (all land, north-south) to a total of 0.65 g/ha and year over one crop rotation. Today each of these sources fills the space. A target for the next five, ten and 15 years for each of the sources should be set. Through purposeful decisions and actions by farmers, authorities, politicians and industry, our society in ten or 15 years can hopefully reach a balance between addition and reduction of cadmium in Swedish soils.

References

Andersson, A., 1991. Trace of elements in agricultural soils. Fluxes, balances and background values. Naturvårdssverket rapport 4077.

Börjesson, I. Cerealia Research Funds, 1995. Personal communication.

Jonsson, A., 1990. Bedömning av spannmålens innehåll av kadmium baserad på SGUs geokemiska kartor. Slutrapport till Stiftelsen Cerealia Research Funds, projekt 128.

Naturvårdsverket, 1993. Renare slam. Åtgärder för kommunala avloppsreningsverk. SNV rapport 4251.

Wikström, L., Skånska Lantmännen, 1995. Personal communication.

Developing an Australian Cadmium Minimisation Strategy

Graeme Evans

Department of Primary Industries and Energy
Australia

General considerations

In developing a national cadmium minimisation strategy, Australia will need to consider and integrate a number of key issues, particularly:

- the extent of the current and future potential health risks, both dietary and environmental;

- impacts on domestic industry;

- impacts on international trade;

- the major sources of cadmium contamination; and

- appropriate mechanisms to minimise human exposure to cadmium.

Present dietary health

The evidence suggests that, by and large, Australian food intake contains cadmium well below Permitted Tolerable Weekly Intake (PTWI) levels. This indicates that cadmium poses little general health risk for the majority of Australians – in the short term, at least, although some segments of the population might be at higher than others. Infants (two to six years old) and those whose diet contains a large amount of ruminant offal and/or crustaceans are obvious examples.

Present environmental health risk

Cadmium levels in Australian soils are generally low, with the contamination normally less than half a microgram per kilogram. However, soils near a number of industrial sites have a much higher cadmium content. For example, a South Australian study showed cadmium levels were up to 164 micrograms per kilogram. It should be emphasised that health risks due to industrial contamination by cadmium in Australia are much less than in many other parts of the world.

Impact on trade – exports

A higher proportion of Australia's economic activity is dependent on international trade than is the case in any other industrialised nation. Australian exports of food and agricultural products have been very little affected by their cadmium content. There have been a few minor problems, usually associated with cadmium levels in offal. However, trade agreements, together with legislation restricting or preventing the sale of some offal for food, have resolved these problems at the relatively low cost of less than $0.5 million per annum.

Impact on trade – imports

Australia's Maximum Permitted Concentration (MPC) standards have been enforced to prevent importation of products with levels of cadmium exceeding our legal limits. A number of products from different parts of the world have a level of cadmium contamination exceeding our MPC standards. Peanuts grown in China are the most obvious example, with a high proportion of such shipments being denied access to Australia.

Cadmium in Australian agriculture

Cadmium causes a few problems for Australian agricultural production. Although Australian soils are very low in cadmium, our crops contain levels of cadmium which are very similar to those of crops grown overseas on soils which may have a much higher cadmium content. This is because Australian soils often have a combination of characteristics quite different from those in many overseas countries. A high proportion of our soils are sandy, highly weathered, acidic, frequently zinc-deficient, and an increasing number are becoming saline. All these factors combine to make cadmium more available to plants and animals than in soils of many other countries containing comparable or even high levels of the element.

Sources of cadmium in Australian agriculture

Cadmium has been added to Australian soils at a much faster rate than its loss through leaching and plant and animal uptake. The main source of Australian soil cadmium is phosphatic fertilizers. Although the level of impurity of the element in many of the high-analysis phosphate fertilizers is much lower than that in superphosphate (previously the standard phosphate fertilizer), they still remain the main source of our soil cadmium.

Industrial emissions, resulting in high levels of cadmium being released into the air, and sewage sludge – both major sources of soil cadmium in many industrialised and heavily populated countries – are of much less significance in Australia. For instance, reliable estimates suggested that, in 1990, sewage contributed only about 1 per cent of the amount of cadmium added to the soil by the fertilizers. Whilst it can confidently be predicted that sewerage will contribute an increasing proportion of the cadmium added to the soil, phosphate fertilizers appear likely to be the main source of contamination, at least for some time.

It is clear that cadmium contamination poses a number of real problems for Australian agriculture and horticulture.

Managing risk of exposure to cadmium

Establishing PTWI standards is the responsibility of a joint WHO/FAO committee (JEFCA). MPC standards in food and agricultural products are determined by individual countries, and the Australian standards are set by the National Food Authority.

Australian standards are amongst the world's most comprehensive and stringent. Many countries either do not legislate for maximum cadmium levels in food or allow a higher level of contamination than we do. Very few have different MPCs for different foods. Some countries, the United Kingdom and Canada for example, rather than regulate the cadmium levels in food, prefer to control the entry of cadmium into the environment as the main strategy to prevent its potentially harmful effects.

Elements of an Australian cadmium minimisation strategy

Australia is well placed to develop a cadmium strategy, being in the vanguard in thinking about the element. Some thought has already been given to the development of a national strategy, and we have a more stringent and comprehensive approach to MPCs that in other countries.

Within the broad field of agriculture, the major source of soil cadmium in Australia has been fertilizers. To reduce the gap between the input and loss of cadmium in the soil, it is obviously important to reduce the level of cadmium contamination in fertilizers. Significant steps have already been taken to achieve this objective, and it can be predicted with some confidence that further reduction in cadmium contamination of phosphate fertilizers will occur.

It is possible to improve the management of cadmium in agriculture through a number of other measures which are effective over differing time scales. Removing zinc deficiency, for example, could be effective within a short time. Other measures, such as changing the composition of the pasture, would obviously take longer, whilst breeding cultivars with a reduced rate of cadmium accumulation could only be a long-term objective.

Key issues needing consideration in developing an Australian strategy should include:

- an assessment of the relative importance of dietary cadmium as a likely health risk;

- the social and economic costs and benefits of maintaining our present stringent MPC standards;

- the undesirability of the current practice of changing MPCs on a case-by-case basis;

- the fact that the development of a national strategy for cadmium, as it relates to issues of health, trade, industry and agriculture, necessitates a wide-ranging review of our current standards, their monitoring and enforcement;

- the recognition that because cadmium, in individual foods and diets, is closely interrelated with health, trade, industry and agriculture, the strategy must be both comprehensive and holistic;

- the roles and responsibilities of the Australian Commonwealth and State governments; and

- an evaluation of the relative cost-effectiveness of regulating dietary cadmium intake through MPC standards on foods compared with improved management of cadmium inputs at their source, in agricultural production for example.

A high-level Government Working Group is being established to develop a national cadmium minimisation strategy for Australia.

Sustainable Cadmium Management in Agriculture: Balancing the Cadmium Fluxes in Arable Land and Grassland

Jens Folke and Lars Landner

European Environmental Research Group

Introduction

There is no question that the current influx of cadmium to the topsoil of agricultural land considerably exceeds the efflux in some regions of the OECD countries (Lander et al., 1995). Although the rate of cadmium accumulation in agricultural soils has decreased over the past few years due to reduction in atmospheric deposition and in fertilizer application rates, there is still an ongoing cadmium accumulation over large areas of food-producing land. Whether or not this accumulation may cause detrimental effects in the general population is a matter of dispute. Nevertheless, our basic food items such as cereals and vegetables, grown on sensitive soils where the cadmium uptake from the soil solution to the crop is particularly efficient, are definitely contaminated with cadmium at levels that have given rise to concern in several countries, particularly in regions where acidification of the soil is a problem, e.g. many Northern European countries (Andersson, 1992a; 1992b).

Irrespective of the controversial question about the relationship between the cadmium accumulation in agricultural soil and the human health risk, the whole issue could be discussed in the light of the principle of sustainable use of our natural resources. It might be claimed that a sustainable use of agricultural land would imply that we have the duty to ensure that the arable soils we leave to coming generations must not be enriched in toxic and potential harmful materials such as cadmium.

This paper is focused on "sustainable cadmium management in agriculture", i.e. what does it take to balance the cadmium fluxes in arable land and grassland. However, the paper does not address the controversial question of whether or how soon a steady-state situation needs to be established.

Fluxes to and from agricultural soils

There is a great variability in the tendencies for long-term accumulation of cadmium in agricultural soils, depending on a great amount of natural factors such as adsorption capacity of soils, agricultural practices, crop profiles, livestock herds, etc. and on atmospheric deposition. Figure 1 illustrates these material fluxes, which are discussed in further detail in Landner et al., 1995.

The real aptitude of a soil to be enriched in cadmium over the long term must be assessed on the basis of specific studies of that soil type, in connection with knowledge about the actual uses of the soil. This is clearly demonstrated in work conducted by Fraters and van

Beurden (1993), who suggest that for about 15 per cent of the agricultural land the cadmium concentration in the soil solution of the topsoil is around or above a value of 1.5 µg/l, and that for a further 7-10 per cent of agricultural land the rate of increase of cadmium in the topsoil exceeds 25 per cent per 100 years.

Qualitative data on the magnitude of these fluxes are scarce, and data are often uncertain estimates rather than true measurements. However, available data from various OECD countries are given in Table 1.

Figure 1

The main pathways of cadmium flux to and from agricultural soils

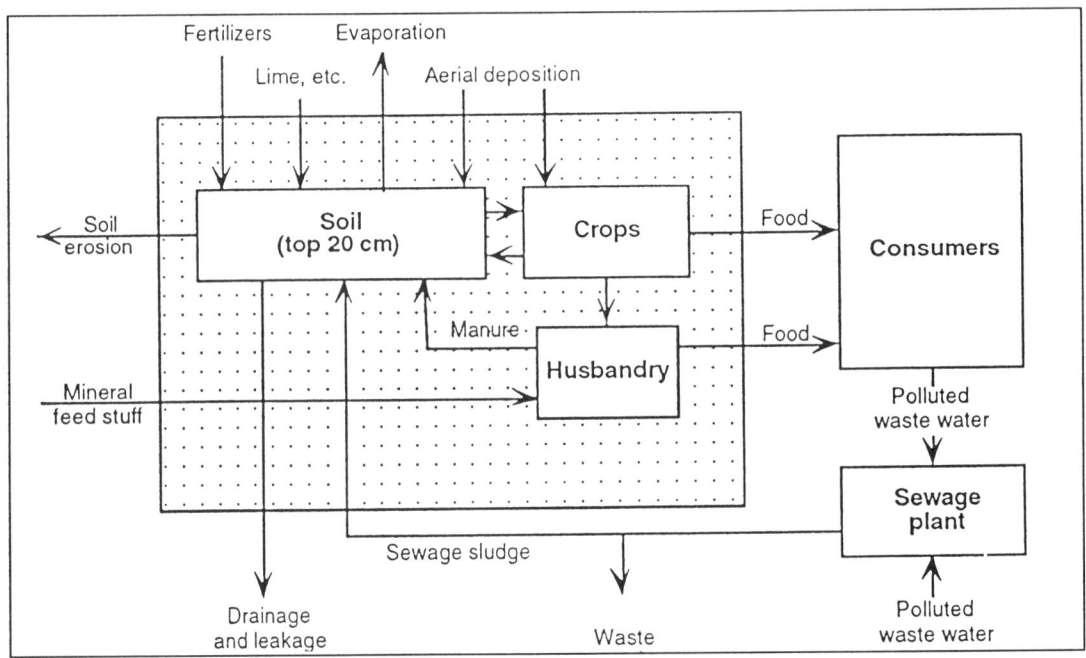

Source: Kongshaug, 1992

How far are we from steady-state?

There is great uncertainty concerning many of the data presented in Table 1. We have changed a few of the older estimates after comparison to other data, as indicated in the notes to the table. But it is still difficult to establish credible mass balances, and the figures in the table should be taken with great precaution. Particularly, the efflux data seem to be erroneous or gross estimates. Holm et al. (1995) estimated at a great variation in the yearly cadmium efflux from 15 different European soils, i.e. from 0.05 to 0.65 per cent of the total soil cadmium per 100 mm net infiltration. Two almost identical soil samples with cadmium soil concentrations of 0.18 mg cadmium/kg (≈450 g cadmium/ha) showed this great variation, i.e. the only significant difference was pH, which was 7.8 for the low leaching sample (0.7 g/ha/year) and 5.1 for the high leaching sample (7.1 g cadmium/ha/year). Depending on different soil parameters, these authors estimated annual leaching rates on the basis of experimental data ranging from 0.5 to 17 g cadmium/ha/year/100 mm net infiltration. This variation by far exceeds the difference in the influx and efflux data given in Table 1.

Also, some of the influx data are difficult to understand, e.g. why is liming such a dominating factor in Denmark but not in Sweden and Finland? How do we account for the fact that biosolids are only applied on small parcels and not on all agricultural land in a country? Therefore, it may be wise to improve the data base in order to be able to estimate more correctly the mass balances for cadmium fluxes as a basis for future decision making.

As shown in the table, the dominating routes of influx are phosphate fertilizers and atmospheric fallout. Biosolids such as sludge and manure can be very dominating local factors, i.e. on land where biosolids are applied as fertilizer. Because the table covers whole countries on average, it gives a false impression of the importance of manure and sludge in those areas where they are applied.

The dominating routes of efflux from agricultural land are leaching and accumulation in crops. Neither is without problems. Leaching can lead to contamination of ground water and thus drinking water, while accumulation in crops leads to human exposure, which is already high compared to the PTWI (Provisional Tolerably Weekly Intake (WHO/FAO), currently 1 µg/kg (body weight/day). Furthermore, the part of cadmium in crops that is not accumulated in humans often finds its way back to agricultural land through applications of sewage sludge on agricultural land. This cycle is similar to the cycle from grazing cattle. Although the cadmium input in manure comes from both atmospheric fallout and feed phosphates, the atmospheric fallout has hitherto been judged to be the most important factor. Therefore, as the fallout is diminished due to flue gas purification, the cadmium content in manure will be reduced as well.

The majority of areas covered in Table 1 are not in balance with regard to the cadmium flux. The cadmium content in Australian rock phosphate is quite high, and so is atmospheric fallout in urban areas. There are no good data available on the cadmium efflux, but there is no question that, due to the high influx rate, Australian agricultural land experiences the highest accumulation in soil among the areas covered by the table. In Denmark, the situation has improved a lot during the last 15 years, primarily due to a considerable reduction in atmospheric fallout. Areas that are close to a steady-state situation include Finland, the southern Netherlands and northern Sweden. However, the size of the actual leaching from these areas is a bit uncertain and needs to verified before a good mass balance can be established.

Table 1

Cadmium mass balances for selected OECD countries

Country, region	PF	Deposition	Liming	FYM	Sludge	Total influx	Crops	Leaching	Totale flux	Total
Australia, rural areas	4^1	< 0.1-0.4	?	?	not applied	4.2	-0.6^2			<3.6
Australia, urban areas	4^1	3	?	?	not^3 applied	7	-0.6^2			<6.4
Finland	0.1	0.4	0.01	0.13		0.64	-0.5	-0.1	-0.6	0.04
Denmark - clay	0.93	0.5	0.4	0.6	0.05	2.55	-0.6	-0.8	-1.4	1.15
Denmark - sand	1.9	0.5	0.4	0.6	0.05	2.55	-0.6	-1.55	-2.15	0.4
Netherlands - North	1.9	3	?	?	0.4	5.3	-0.6	-2.5	-3.1	2.2
Netherlands - South	1.9	0.9	?	?	0.4	3.2	-0.6	-2.5	-3.1	0.1
Sweden, GSS, 13 kg P/ha/year	0.85	1.1	0.02	0^4		1.97	-0.64	-0.6^5	-1.24	0.73
Sweden, GMB, GNS, SS, 11 kg P/ha/year	0.72	0.78	0.02	0^4		1.52	-0.21	-0.6^5	-0.81	0.71
Sweden, GMB, GNS, 8 kg P/ha/year	0.52	0.78	0.02	0.65		1.97	-0.67	-0.6^5	-1.27	0.70
Sweden, SS, 8 kg P/ha/year	0.52	0.78	0.02	0.35^6		1.67	-0.67	-0.6^5	-1.27	0.40
Sweden, GSK, 8 kg P/ha/year	0.52	0.9	0.02	0.65		2.09	-0.49	-0.6^5	-1.09	1.00
Sweden, N, 6 kg P/ha/year	0.39	0.4	0.02	0.2^6		1.01	-0.35	-0.6^5	-0.95	0.06

Unit is g/ha/year.
GMB: Southeast Götaland.
GNS: Plain district, northern Götaland.
N: Northern Sweden, Norrland.
GSK: Forest districts, Götaland.
GSS: Plain district, south Götaland.
SS: Plain district, Svealand.
1.: Concentration ranges are 413 ± 40 mg Cd/kg P. Assumed rate is 10 kg P/ha/year.
2.: MFG estimate.
3.: The potential is 2-4 t/y for all Australia, to be applied on land near cities. McLaughlin et al., 1995.
4.: No animal husbandry.
5.: Original figure published was -0.06. However, comparison to Danish (clay) and Dutch figures suggests the true figure is -0.6 or higher.
6.: Original figure published was -0.65. However, Cd atmospheric outfall decreases going north, so FYM must contain less Cd also.

Source: MFG estimates as well as Andersson, 1992b; Landner et al., 1995.

How could we reach a steady-state situation?

If the overall objective with regard to cadmium is to achieve a long-term input and output equilibrium of the cadmium content in agricultural topsoil, several strategies must be combined. As stated above, increasing the efflux rate in order to achieve a steady-state level is hardly acceptable unless the purpose is to clean up a contaminated site, so steady-state has to be achieved by reducing the influx. The efflux rate in areas close to steady-state is 1 ± 0.5 mg Cd/ha/year provided soil pH > 7, total soil Cd < ≈ 0.3 mg/kg (≈750 g Cd/ha) and no other extremes apply, so the influx rate has to be reduced to that level to reach the goal.

In urban and industrialized areas the atmospheric fallout can be reduced by flue-gas purification to a level ≈ 0.5 ± 0.1 mg Cd/ha/year. In rural areas the figure is closer to ≈ 0.1 mg Cd/ha/year.[5] Consequently, the sum of influx from phosphate fertilizer, liming, manure and biosolids will have to be reduced to 0.5 ± 0.4 mg Cd/ha/year in urban or industrialized areas and 0.9 ± 0.5 mg Cd/ha/year in rural areas. While the goal in rural areas for cadmium exposure through phosphate fertilizer, liming, manure and sludge may be reached with a reasonable effort that may well include some technical measure of purification of phosphate rock and lime, the goal for urban areas is much harder to reach. The qualities of phosphate fertilizer with the absolute lowest levels of cadmium contamination will have to be used in those areas by selective purchases, and stringent limits will have to be imposed on the cadmium content in sludge and manure to reach the goal. Another important measure is reductions in phosphate application rates to soils. In Denmark the application rate of commercial phosphate fertilizers has gone down from 400,000 tonnes per year in 1989/90 to 315,000 tonnes per year in 1994/95 (Ludvig, 1995). Fields where crops with high requirements for phosphorous are grown may only reach a steady-state if the field is shielded from atmospheric fallout (greenhouses).

In conclusion, it appears that even if atmospheric deposition of cadmium in the whole OECD area could be brought down to the levels typical for Finland, northern Sweden and rural areas of Australia and Canada, there would still be a need to take measures against the cadmium influx with phosphate fertilizers if steady-state is to be reached. Technical measures to reduce the natural cadmium content in fertilizers as well as selective purchases of low level cadmium fertilizers will have to be considered.

---

[5] Hardly any areas in continental Europe meet this definition of a rural area, but areas in e.g. Ireland, Scotland, Canada, New Zealand, etc. may well be covered.

References

Andersson, A., 1992a. "Cadmium in Swedish Soils and Wheat Production." ISTERTH Third International Conference and NTES Fourth Nordic Conference on Trace Elements in Health and Disease, May 25-19, 1992, Swedish University of Agricultural Sciences, Uppsala, Sweden.

Andersson, A., 1992b. "Trace Elements in Agricultural Soils - Fluxes, Balances and Background Values." Report 4077, Swedish Environmental Protection Agency, Stockholm, Sweden.

Fraters, D. and A.U.C.J. van Beurden, September 1993. "Cadmium Mobility and Accumulation in Soils of the European Communities." GLOBE-Europe. Report N°481 505005, National Institute of Public Health and Environmental Protection, Bilthoven, The Netherlands.

Holm, P., T.H. Christensen, S.E. Lorenz, R.E. Hamon, H.C. Domingues, E.M. Sequeira and S.P. McGrath, 1995. "Measured Soil Water Concentrations of Cadmium and Zinc in Plant Pots and Estimated Leaching Outflows from Contaminated Soils." In: Speciation of Dissolved Cadmium in the Terrestric Environment, Peter Holm, PhD Thesis (in Danish, manuscript in English). Institute of Environmental Technology, The Technical University of Denmark, Lyngby, Denmark.

Kongshaug, G., 1992. "Trace Elements in Phosphate Rocks. Problems with Build Up in Soil - Fact of Fiction?" Paper for Fertilizer Industry Round Table, Oct. 26-28, 1992, Baltimore, Md.

Landner, L., J. Folke, M.O.Öberg and M. Aringberg-Laanatzr, 14 September 1995. "Cadmium in Fertilizers - Preparation of an OECD Workshop." Final draft. European Environmental Research Group, Stockholm, Sweden, commissioned by Kemikalieinspektionen, Stockholm, Sweden.

Ludvig, E.T., 1995. "Less Use of Commercial Fertilizer" (in Danish). *Berlingske Tidende* 28 August. Erhvervssektionen, page 1.

McLaughlin, M.J., K.G. Tiller, R. Naidu and D.P. Stevens, 1995. "The Behaviour and Environmental Impact of Contaminants in Fertilizers." *Aust. J. Soil Research*: Submitted.

Aspects of Cadmium Accumulation in Agriculture

Simon W. Moolenaar, Panos Hatziotis and Theo M. Lexmond

Wageningen Agricultural University
Department of Soil Science and Plant Nutrition
The Netherlands

Abstract

Since accumulation of cadmium in soil can damage the quality of agricultural produce when certain concentration levels are exceeded, control of cadmium fluxes is a prerequisite for sustainable agricultural production. At the same time, the ecological functioning of the soil should not be damaged and emissions from the soil should not adversely affect other environmental compartments. For this purpose Dutch policy aims at balancing cadmium input via fertilizers with cadmium output via produce and input via atmospheric deposition, with acceptable leaching losses. To guard these criteria, quality standards for soil, groundwater and agricultural produce have been developed. In order to satisfy these limits a farmer can pay attention to the following aspects:

- the selection of fertilizers with the lowest cadmium and the highest phosphorous content, in order to achieve the lowest possible Cd/P_2O_5 ratio for the inputs;

- the selection of the most appropiate farming system;

- the selection of cultivated crops with pronounced cadmium removal (within accepted levels).

Cadmium balances are a useful tool to appreciate the relevance of cadmium fluxes, the resulting accumulation, and associated risks in agroecosystems. We established the annual cadmium balance in three arable farming systems on marine clay. These systems were a conventional one with the use of either exclusively mineral fertilizers or a combination of organic and mineral fertilizers, an integrated one, and an ecological system.

The annual cadmium balance was expressed as the difference between input and output (I-O) and also as the ratio of inputs over outputs (I/O). This ratio varied from 2.5 (integrated system) via 3 (conventional system with organic fertilizer) and 5.2 (ecological system) to 7.4 (conventional system with mineral fertilizers only). These differences are mainly due to the different levels of cadmium input via fertilizer applications (use of triple superphosphate) and the cultivation of crops that have a high cadmium offtake (carrots, sugar beets, ware potatoes and onions). The ecological system has a higher input/output ratio, since 47 per cent of the total surface is cultivated with grain crops (limited cadmium offtake) and since the inputs via manure applications are significant (partly because of high manure applications for raising the phosphorus status of the soil).

This case of three different agro-ecosystems shows that atmospheric deposition, the selection of the crops, and the selection of the fertilizers have a direct influence on both the annual cadmium balance and the long-term development of cadmium concentrations in soil, groundwater and crops of the farming systems.

The balance concept

The balance of a metal in soil is the result of input minus output. Sustainability requires that, in the long term, input should not exceed acceptable output through removal in crops and leaching (Furrer, 1984; Ferdinandus et al., 1989).

For cadmium balances three aspects are important:

1) Input and output flows

Metal fluxes in agro-ecosystems involve many processes and consequently depend on many factors. Input flows are related partly to external sources that are difficult or impossible to control by the individual farmer (atmospheric deposition, sedimentation during inundation) and partly to purchased materials (fertilizers, manure, feed concentrates, forage, soil conditioners, pesticides). Output proceeds via marketing of produce and losses to the agro-ecosystem's external environment. Output via produce depends strongly on the farming system, output from dairy farms (via milk and meat) being notably smaller than that from arable farms (Van Driel and Smilde, 1990). Losses to the external environment are mainly due to transport in water and are therefore related to metal concentrations in the soil solution and in groundwater. These relationships can be very complicated, as they involve soil properties, hydrological conditions, climate, and farm management.

2) The reliability of data from literature and measurements

Literature data often show large variations, so that there is a need to improve the quality and quantity of the needed data to be able to improve on estimates and best guesses.

3) The judgement of the balance in view of sustainability

In principle, the different flows can be measured or calculated and an annual balance can be estimated. In practice, there is a problem in giving a "full picture", mainly because of uncertainty concerning leaching.

At the moment, the balance concept has been developed with respect to diffuse contamination. The concept is simple: soil (e.g. the unsaturated zone or the ploughed layer) may be considered as an environmental compartment (a grey or black box) that may be characterized by inflows and outflows related to the accumulated amount in the soil compartment.

Referring for some more details of the derivation to Boekhold and Van der Zee (1991), the soil compartment balance equation is

$$\frac{d}{dt}(G) = A - BG^m - CG^{1/n} \tag{1}$$

where the accumulation rate in soil equals the input rate (A, which is independent of G and assumed constant) minus removal rates (i) in harvested product (i.e. BG^m) and (ii) by leaching (i.e. $CG^{1/n}$). The powers m and n depend on whether or not linear relationships are assumed. G is the total content of cadmium in soil. The rate parameters A, B and C can be measured and/or estimated.

The plant uptake rate coefficient (B) and the leaching rate parameter (C) are in this illustration assumed to be constant in time. Other factors of interest can be incorporated in this balance as well. With the balance (1) agro-ecosystems can be screened and compared with respect to their crop uptake, leaching and accumulation in soil relative to critical limits. This approach will be illustrated in the following case study on the annual cadmium balance at the Nagele experimental farm in the Netherlands (for further details, see Hatziotis, 1995).

Cadmium balances at the Nagele experimental farm

The annual balance of cadmium might be a useful indicator to determine the sustainability of a farming system. This case study established the annual cadmium balance in three arable farming systems: conventional arable farming system (CAFS), integrated arable farming system (IAFS) and ecological arable farming system (EAFS), practised at the Nagele experimental farm (as described in Vereijken, 1992) in 1994. The conventional system is divided into two subsystems with the use of organic fertilizers (CAFS OF), and with the exclusive use of mineral fertilizers (CAFS MF). The soil is marine clay with 24 per cent clay, 2.6 per cent organic matter, 10 per cent $CaCO_3$ and pH-KCL at 7.4. The annual cadmium balance has been determined from annual inputs, outputs and losses. The different systems have been compared as to the mechanisms that govern these fluxes.

The mean total cadmium content (Cd-T) of the soil was 0.5 mg/kg for all three farming systems, indicating background contamination. Cadmium intensity (Cd in 0.01M CaC_{12}), an approximation of the bioavailability of cadmium, was not detectable. Therefore, it was assumed to be less than 0.024 microgram per litre (half the detection limit of GFAAS). High pH, and $CaCO_3$ content, were the key factors for the low cadmium intensity.

Cadmium inputs were determined from the atmospheric deposition (assumed constant at 2 g/ha.yr) and the fertilizer applications. The mineral fertilizers CAN and K-60 were practically free of cadmium. Triple superphosphate (TSP) had a cadmium content of 31 mg/kg with a Cd/P_2O_5 ratio of 68 mg/kg. The three organic fertilizers (goat, cattle and poultry manure) had comparable contents of cadmium. The Cd/P_2O_5 ratio of liquid poultry manure was the lowest (7.7) while for goat and cattle manure the corresponding ratios were 30.6 and 20.7 mg/kg, respectively. For the ecological and integrated systems the contribution of atmospheric deposition to the total cadmium inputs was higher than the contribution of the manure applications.

Tissue concentrations of cadmium for all the crops were lower than the Dutch critical levels. Celeriac had the highest and grains the lowest cadmium tissue concentrations among all crops. Bean, carrot, chicory and sugar beet also had high tissue concentrations of cadmium. The cadmium offtake was within the range of 0.01-2 (g/ha.yr). This implies that the cadmium outputs cannot level the inputs via cadmium atmospheric deposition, except in cases where crops with high cadmium adsorption, such as celeriac, are cultivated. Further, there was no increase in the cadmium uptake by the crops in the fields with high cadmium inputs. This can be explained by the high adsorption capacity of the particular soil.

Cadmium losses via leaching were estimated from the cadmium concentration in 0.01 M CaC_{12} and the net precipitation (0.2 m^3/m^2.yr). Cadmium leaching was limited (less than 0.06 g/ha.yr) due to the low cadmium solubility.

The annual cadmium balance was expressed as the difference between input and output (I-O in g/ha) and also as the ratio of inputs over outputs (I/O). The integrated system (I-O = 1.5; I/O = 2.5) and conventional system with organic fertilizer (I-O = 1.7; I/O = 3) had supremacy over the ecological system (I-O = 2.8; I/O = 5.2) and conventional system with mineral fertilizers only (I-O = 5.7; I/O = 7.4), due to the relatively low level of cadmium input via fertilizer applications (no use of triple superphosphate) and the cultivation of crops that have a high cadmium offtake (carrots, sugar beets, ware potatoes and onions).

The ecological system has a higher input/output ratio, since 47 per cent of the total surface is cultivated with grain crops (limited cadmium offtake) and since the inputs via manure applications are significant. It has to be noticed (for a fair comparison) that this is partly due to high manure applications for raising the phosphorus status of the soil. Within the conventional systems the cadmium balance shows a much larger discrepancy for the system with mineral fertilizers only. The triple superphosphate applications are in this case a more important factor than the cadmium atmospheric deposition.

Consequently, the selection of the crops, and the selection of the fertilizers, have a direct influence on the cadmium balance of the farming system.

Cadmium management related to phosphorus

All phosphorous fertilizers contain cadmium, and therefore there is an interconnection between cadmium and phosphorous flows in agro-ecosystems. Sustainable cadmium management implies there are two main criteria:

1) additions of cadmium to the soil, via atmospheric deposition and fertilizer applications, should not exceed the **acceptable** output of cadmium by harvest and leaching;

2) the total cadmium content in soil should not exceed the relevant limit value.

In fact, these two criteria are the general indicators for soil protection according to the Dutch environmental policy for cadmium, which aims at balancing cadmium input via fertilizers with cadmium output via produce and input via atmospheric deposition, with acceptable leaching losses (Anonymous, 1991).

To guard these criteria, quality standards for soil, groundwater and agricultural produce have been developed, and in order to satisfy these limits a farmer can pay attention to the following aspects:

- the selection of fertilizers with the lowest cadmium and the highest phosphorous content, in order to achieve the lowest possible Cd/P_2O_5 ratio for the inputs;

- the selection of the most appropiate farming system;

- the selection of cultivated crops with pronounced cadmium removal (within critical levels).

These points have to be handled carefully. For example, we can argue that the increase in production of bean, sugar beet or carrot, with a relevant decrease in grain crops production, can improve the soil quality (limited to cadmium!). However, this might have consequences for cadmium intake (public health) if consumption patterns change too.

Mineral fertilizers

Regarding the inputs of cadmium in relation to phosphorous application of fertilizers, mineral phosphorous fertilizers such as phosphates (ordinary and triple superphosphates) can have a very high content of cadmium. Thus, when aiming for minimisation of cadmium inputs to agricultural soils, phosphorous mineral fertilizers should either be avoided in the fertilization strategy or qualitatively improved in terms of Cd/P_2O_5 ratio. Consequently, a sustainable steady-state of cadmium could be achieved sooner and at lower levels of cadmium output and total accumulation.

Alternative sources of phosphorous are organic fertilizers, either manures or composts.

Manures

The results for the manure samples at Nagele show variable ratios of Cd/P_2O_5. This variability confirms that there is no fixed ratio between cadmium and phosphorous in manures. Among the examined manures, poultry manure had the lowest Cd/P_2O_5 ratio. Before the selection of the most appropiate manure, there is always the need for determination of the Cd/P_2O_5 ratio.

Composts

Composts are a second group of organic fertilizers. In Holland much debate is going on about the safety of compost applications from a heavy metals contamination point of view.

The two main groups of composts are municipal/urban composts and agricultural composts. The quality of each is quite different. In general, municipal composts have the lowest quality because of the likelihood of contamination in the urban environment. Two main factors are related to the low quality of urban composts:

1) the inclusion of non-organic materials or contaminants in the compost;

2) the high contribution of urban soil in the final product, which is in most cases contaminated with heavy metals.

An example illustrating the importance of the second factor is the Source Separated Organics compost (SSO-compost), known in Holland as VFG-compost (compost from Vegetables, Fruit and Garden waste). The final material produced from this type of compost contains on average 70 per cent of soil (on a dry weight basis). Depending on local circumstances, the agricultural composts may have the best quality.

In judging these different types of fertilizers, one should be careful to take into account all relevant aspects which are related to the use of these fertilizers.

Comparing mineral fertilizers, manure and composts from the point of view of nutritive value will probably show most composts are not at all preferable, since they have the highest cadmium/P_2O_5 ratio. A calculation for SSO compost with optimistic values of 0.7 mg/kg (d.m.) cadmium and 0.9 per cent P_2O_5 shows that the cadmium/P_2O_5 ratio is 77.8, being even higher than the ratio of triple superphosphate (68.3). However, if the target of the fertilization strategy is mainly the maintenance of soil organic matter and nutrients supply is only of secondary importance (e.g. for areas saturated with phosphorous), then composts can be advantageous or even necessary due to their high organic matter content.

Furthermore, Moolenaar et al. (1995) have shown that to justify future regulations with regard to fertilizer quality and application it is necessary to deal with several complications with respect to the correct way of calculating resulting heavy metal accumulation. These complications result from the fertilizer composition and the resulting change in soil composition. This is elaborated upon in the Appendix (Fertilizer and Soil Composition).

Long-term cadmium accumulation

The long-term behaviour of cadmium was predicted for the ecological system according to the balance equation (1) in fields with barley (minimum cadmium offtake) and celeriac (maximum cadmium offtake). The input (parameter A) is considered to remain constant. Crop uptake of cadmium is considered to be linearly related to the soil cadmium content. Hence the value of power m (eq. 1) is equal to one.

The results of these calculations (Appendix: Long-term Accumulation) show a big difference between the barley and celeriac fields. For the barley field, a steady-state condition will be achieved after approximately 9000 years. At that time the total cadmium content of the soil will be 2.06 mg/kg, corresponding with a contaminated soil, since the

Dutch reference value for this soil is 0.62 mg/kg. Also the part of the cadmium inputs (2.53 g/ha) that will be adsorbed by the crops (0.19 g/ha) and the part that will be lost by leaching (2.34 g/ha) can be calculated. Assuming that there will be no depression of yield, the cadmium tissue concentration (cp) will be 0.048 mg/kg of fresh material. This is much lower than the Dutch critical level for grains (0.15 mg/kg fresh material). Leaching will result in 0.78 Tg/l, thus doubling the Dutch target value of 0.4 Tg/l for the cadmium concentration in groundwater.

The corresponding values for the celeriac field, after a steady-state has been achieved (within 1000 years), are a total cadmium content of 0.58 mg/kg, crop output of 2.5 g/ha, losses by leaching of 0.05 g/ha, a cadmium tissue concentration of 0.0 mg/kg of fresh material, and a cadmium concentration in groundwater of 0.02 Tg/l. In this latter case all parameters are below the Dutch reference values (Appendix: Reference Values).

The comparison shows that the difference of these scenarios is due to the plant uptake factor K_{up}, which is minimum for barley and maximum for celeriac.

The same procedure can be applied not only for every field (with corresponding crop), but also for every farming system, using the mean area weighted values of the parameters A, B and C in each entire farming system (Appendix: Long-term Accumulation).

In all farming systems the total cadmium content (in the steady-state) will exceed the Dutch reference value, while the cadmium concentration in groundwater is below the target value except for the conventional system with mineral fertilizer, where this value will be marginally exceeded. The cadmium concentration in products is generally below the critical levels. Among the grain crops, winter wheat exceeds the critical level except in the integrated system. The summer wheat in the ecological system also exceeds the critical level. For all the other crops, only the bean and celeriac in the ecological system and sugar beet and chicory in the conventional system with mineral fertilizer exceed the critical levels.

Following the same procedure for the calculations, we illustrate the probable improvement of cadmium accumulation in the soil of the conventional system with mineral fertilizer for two different scenarios. In the first scenario (CAFS MF-fert.) the cadmium content in TSP is cut back to 15 mg/kg P_2O_5 (possible future content). In the second scenario (CAFS MF-dep.) the same reduction in fertilizer content is combined with a reduction in atmospheric deposition to 1 g/ha.yr, which is the Dutch target value (Appendix: Long-term Accumulation).

Based on the data (partly presented in the Appendix), we can conclude that:

- The total cadmium content is exceeding the Dutch reference value in all the actual farming systems. The integrated system has the lowest and the conventional system with mineral fertilizer the highest total content.

- Even the hypothetical (improved) scenarios for fertilizer cadmium content and deposition do not result in a total cadmium concentration which is lower than the Dutch reference value.

- The cadmium concentration in groundwater is below the critical level for all the farming systems, except the conventional system with mineral fertilizer.

- Among the grain crops, winter wheat exceeds the critical level except for the integrated system and the conventional system "adjusted" for fertilizer and deposition input. The summer wheat in the ecological system also exceeds the critical level. For all the other crops, only the bean and celeriac in the ecological system, and sugar beet and chicory in the conventional system with mineral fertilizer, exceed the critical level.

- Although the total cadmium contents of the farming systems are rather high, c and c_p do not remarkably increase due to the high adsorption capacity of this particular soil. These findings are highly dependent on the soil's buffer capacity. On a sandy soil (with a low pH) it is expected that cadmium concentrations will increase more in groundwater than in crops (see e.g. Boekhold and Van der Zee, 1991).

- The results show that it is possible to judge the (long-term) sustainability of different agroecosystems. A combined effort of reducing cadmium input via fertilizers and atmospheric deposition is necessary to reach a sustainable steady situation (all concentrations below the critical levels).

Acknowledgements

This work was partly funded by a grant from the Dutch Organization for Scientific Research (NWO), Priority Program Sustainability and Environmental Quality.

References

Anonymous. 1991. Cadmiumbeleid. Tweede Kamer, 1990-1991, 22 197, nr. 1.

Boekhold, A.E., S.E.A.T.M. van der Zee. 1991. Long-term effects of soil heterogeneity on cadmium behaviour in soil. J. Contam. Hydrol. 7: 371-390.

Commission of the European Communities. 1986. Council Directive (86/278/EEC) on the protection of the environment, and in particular of the soil, when sewage sludge is used in agriculture. Off. J. European Community. Annex 1a: 6-12.

DOOF. 1991. Decree on the Quality and Use of Other Organic Fertilizers (Besluit kwaliteit en gebruik overige organische meststoffen). Staatsbl. 1991: 613.

Driel, W. van, K.W. Smilde. 1990. Micronutrients and heavy metals in Dutch agriculture. Fert. Res. 25: 115-126.

Ferdinandus, H.N.M., Th.M. Lexmond, F.A.M. de Haan. 1989. Heavy metal balance sheets as criteria for the sustainability of current agricultural practices (in Dutch with English summary). Milieu 4: 48-54.

Furrer, O.J. 1984. Konzept zur Festlegung von Grenzwerten für Schwermetallimmissionen in den Boden. Schweiz. Landwirtsch. Forsch. 23: 195-199.

Hatziotis, P. 1995. Annual Cadmium Balance and Sustainable Land Use: A comparative study of ecological, integrated, and conventional farming systems. MSc thesis, WAU.

Lexmond, Th.M., Th. Edelman, W. van Driel. 1986. Voorlopige referentiewaarden en huidige achtergrondgehalten voor een aantal zware metalen en arseen in de bovengrond van natuurterreinen en landbouwgronden. Bijdrage naar aanleiding van de Discussienotitie Bodemkwaliteit. Verslagen en Mededelingen 1986-2. Vakgroep Bodemkunde en Plantevoeding, LUW.

Moolenaar, S.W., F.A.M. de Haan, Th.M. Lexmond, S.E.A.T.M. van der Zee. 1995. Accumulation of heavy metals in soil: The sustainability approach. In: Proceedings of the Third International Conference on the Biogeochemistry of Trace Elements. Paris, May 15-19, 1995. Submitted.

Vereijken, P. 1992. A methodic way to more sustainable farming systems. Netherlands Journal of Agricultural Science 40: 209-223.

Wit, A. de. 1989. De overdracht van zware metalen van grond naar gewas: het effect van zinkcompetitie op de opname van cadmium door spinazie (Deel 1). MSc thesis, Department of Soil Science and Plant Nutrition, WAU, Wageningen.

Appendix

Fertilizer and Soil Composition

Not only the reference values (in the Dutch situation), but also the heavy metal contents themselves depend on the clay and organic matter content of the soil This implies that it is important to make an accurate calculation of organic matter, clay and heavy metal contents in the course of time. Using a case study of SSO compost the main point of this so-called sustainability approach was stressed, namely the time dependency of the soil bulk density (rho), clay (L) and organic matter (H) contents. These contents change significantly within 100 years: a time scale relevant to the sustainability principle (Moolenaar et al., 1995).

So, when calculating long-term accumulation, it is also important to take into account the resulting (long-term) changes in soil composition. The use of mineral fertilizers does not result in an extra input of soil particles and organic matter. The example for SSO compost (70 per cent d.m. soil particles) showed that due to regular SSO compost applications, the contents of clay and organic matter will change. This influences the calculation of the cadmium contents in the course of time considerably (Moolenaar et al., 1995).

These aspects should also be taken into account in the comparative evaluation of the long-term cadmium effect of compost versus mineral fertilizers and animal (e.g. pig) manure.

Long-term Accumulation

Leaching of cadmium can be described according to the Freundlich sorption equation ($q = K.c^n$), in which k and n are affinity constants specified according to the soil type. Parameters q and c stand for soil content and concentration in soil solution, respectively. The value of (dimensionless) parameter n can be taken as 0.5 for the soil at Nagele. This results in the following form of the balance equation (2):

$$\frac{d}{g}(G) = A - BG - CG^2 \qquad (2)$$

in which:

    A = (constant) input rate

    $B = -K_{up} = Y.c_p/l * 1/G$

        in which K_{up} is the constant uptake rate (yr-1) for Cd by a crop and

        Y: yield of product (kg/m².yr)

c_p: tissue concentration of Cd in dry matter (Tmol/kg)

l: thickness of plough layer (0.3 m)

$$C = -(\theta \cdot v/l) \cdot (1/rho.k)^2$$

in which

θ: volumetric water content of the soil (0.3 m^3/m^3)

v: interstitial flow velocity (1 m/yr)

k: affinity constant (6.96 10^{-6} mol0.5.kg-1.m1.5)

rho: dry bulk density of soil (1400 kg/m^3)

Equation (2) can be solved analytically. According to Boekhold and Van der Zee (1991) G then equals:

$$G = \frac{(B + \sqrt{D})(B-\sqrt{D})(e^{\sqrt{D}t} - 1)}{2C[B + \sqrt{D} - (B - \sqrt{D})e^{\sqrt{D}t}]} \quad \text{with } D = B^2 - 4AC \text{ and } G_{t \to \infty} \frac{B + \sqrt{D}}{-2C}$$

The A, B and C parameters equal 7.5 (Tmol/m^3.yr), -0.22*10^{-4} (1/yr) and -1.05*10^{-8} (m^3/Tmol.yr) respectively for the barley field and 7.6 (Tmol/m^3.yr), -19.8*10^{-4} (1/yr) and -1.05*10^{-8} (m^3/Tmol.yr) respectively for the celeriac field.

The parameters A, B and C of the different scenarios take the following values:

A_{EAFS} = 10.2 Tmol/m^3.yr

A_{IAFS} = 7.41 Tmol/m^3.yr

$A_{CAFS\ OF}$ = 7.47 Tmol/m^3.yr

$A_{CAFS\ MF}$ = 19.35 Tmol/m^3.yr

$A_{CAFS\ MF-fert.}$ = 8.87 Tmol/m^3.yr

$A_{CAFS\ MF-dep.}$ = 5.93 Tmol/m^3.yr

B_{EAFS} = -6.34×10^{-4} 1/yr

B_{IAFS} = -10.71×10^{-4} 1/yr

$B_{CAFS\ OF}$ = -8.03×10^{-4} 1/yr

$B_{CAFS\ MF}$ = -8.43×10^{-4} 1/yr

C   = -1.05×10^{-8} m3/Tmol.yr

The total soil content (G) is the sum of the fixed and the adsorbed amount of cadmium. The fixed fraction of cadmium cannot be released in the soil solution, and thus it is more realistic to exclude it from the calculation of leaching and uptake. De Wit (1989) estimated the fixed fraction for the Nagele soil to be 0.28 mg/kg. The "corrected" total soil content has been used to calculate the total soil content, leaching and plant uptake of the farming systems according to the analytical solution of (2). Also the corresponding values for cp and c can be calculated, using average uptake rate (kup), yield and G (at steady-state) values of each farming system. In this way it is possible to get an overall picture for the cp, c and G of every farming system when a steady-state condition is achieved (Table 1).

The results for the scenarios in which cadmium input by fertilizers has been reduced from 68 to 15 mg cadmium/kg P_2O_5 (CAFS MF-fert.) and in which the fertilizer input as well as the input by deposition have been reduced (CAFS MF-dep.) are shown, too (Table 1).

The calculated steady-state concentrations for the different systems and scenarios can be found in Table 1.

Reference Values

It is unsatisfactory to translate "good soil quality" into total contents, because many soil properties are related to e.g. bioavailability of heavy metals in a complicated way. Although the specific (chemical) behaviour in soil is not taken into account when only total contents are regarded, soil quality standards are generally expressed in (total) contents. Hence, it is important to know when soil contents exceed these standards.

In the Netherlands the soil reference values have a built-in soil type correction. Consequently, the accepted burden is about equal for different soil types, using this soil type correction. These reference values for metal concentrations in soils are derived from a study of the concentrations of metals found in nature reserves in the Netherlands (Lexmond et al., 1986), which, although not in urban or industrial zones, will have been impacted by atmospheric deposition of metals. They are linearly related with the weighed sum of clay (L) and organic matter (H) content (Table 1). In the EC Directive on the use of sewage sludge on agricultural soils there is no differentiation with regard to soil type at all. In fact, the accepted burden appears to be largest for the most vulnerable soils (i.e. with the lowest soil buffer capacity and the lowest initial heavy metal contents). The Dutch Decree on the quality and use of Other Organic Fertilisers (DOOF, 1991) came into force implementing this EC Directive on the use of sewage sludge on agricultural soils. The Dutch reference values serve as limit values in this decree (Table 2).

Table 1

Steady-state concentrations in soil, crop and groundwater for different farming systems

	Reference Values	EAFSS	IAFS	CAFS OF	CAFS MS	CAFS MS fert. 15 mg/kg	CAFS MS dep. 1 g/ha
G (mg/kg)	0.62	**1.34**	0.81	0.95	**1.78**	1.03	0.8
C (µg/l)	0.40	0.21	0.05	0.08	**0.41**	0.1	0.05
C_n	Critical levels						
C_p carrot	0.20	0.10	0.05				
C_p seed potato	0.10	0.05	0.02	0.03	0.06	0.03	0.02
C_p ware potato	0.10		0.03	0.04	0.10	0.05	0.03
C_p s. barley	0.15	0.02		0.01	0.03	0.015	0.01
C_p s. wheat	0.15	**0.32**					
C_p w. wheat	0.15	**0.28**	0.14	**0.18**	**0.4**	**0.2**	0.14
C_p bean	0.10	**0.19**					
C_p oat grains	0.15	0.15					
C_p onion	0.10	0.04	0.02	0.02	0.06	0.03	0.02
C_p celeriac	0.10	0.25					
C_p sugar beet	0.10		0.06	0.08	**0.18**	0.09	0.06
C_p chicory	0.20			0.13	**0.28**	0.15	0.1

Table 2

Maximum permitted values for metal concentrations

	EC	Netherlands
Zinc	150-300	$50+1.5*(2LL+H)$
Cadmium	1-3	$0.4 + 0.007*(L+3H)$
Copper	150-400	$15 + 0.6*(L+H)$
Lead	50-300	$50 + L + H$

(Commission of the European Communities Directive 86/278/EEC, 1986; DOOF, 1991)

Cadmium Accumulation in the Soil – An Increasing Problem

Kierstin Petersson Grawé

National Food Administration
Uppsala, Sweden

The National Food Administration has the national responsibility for the safety of food in Sweden. Concern about dietary cadmium exposure in the general population is therefore of major interest to us.

Trends

In 1988 it was estimated that 65 per cent of the cumulative world cadmium production had taken place during the last two decades (ICPS, 1992). Production and use have increased since then, according to the OECD Risk Reduction Monograph on cadmium. In the same OECD publication, increasing levels of cadmium in soil in several countries, e.g. Germany, the Netherlands, Norway, Finland and Sweden, are reported. It is thus obvious that the magnitude of the cadmium problem will increase unless steps are taken to reduce the emissions of cadmium. Generally the efficiency of such steps increases with the nearness to the source of the emission.

It is estimated that if the cadmium accumulation in the Swedish soil continues at the same rate as today, there will be a 50 per cent increase in cadmium content in winter wheat in Sweden 90 years from now (Gerhardsson et al., 1994). This accumulation is mainly due to atmospheric deposition and the use of commercial phosphor fertilizers. It is thus a very slow process. A significant increase in dietary cadmium intake in the general population can probably not be expected within the next few decades.

It is generally accepted that the body burden of cadmium in the general population increases in cases of heavy environmental cadmium pollution. In addition, renal function is affected in the general population in these cases. Recent reports from Belgium (Buchet et al.,1990, Staessen et al., 1994) indicate renal effects in the general population at substantially lower exposure levels. In earlier studies, i.e. the Shipham and Stolberg studies (Strehlow and Barltorp, 1988, Ewers et al., 1985) increased body burden was reported but not renal effects.

In the above mentioned Cadmibel study by Buchet and co-workers, renal effects were observed in the general population although the provisional tolerable weekly intake (PTWI) of 400-500 µg cadmium, corresponding to 60-70 µg/daily set by Joint FAO/WHO Expert Committee on Food Additives and Food Contaminants (JECFA) in 1993, was not exceeded. Two recent reports confirm these results (Roels et al., 1993, Järup and Elinder, 1994).

The options of an authority responsible for food safety

The exposure to some heavy metal contaminants in food, e.g. lead and methylmercury, can relatively easily be avoided or at least decreased by reducing the consumption of a small number of foodstuffs. The management of risk reduction can in part be handled by setting limit values for lead and methylmercury in certain food items, together with an effective control. Recommendations to the general public on restricted consumption of certain foodstuffs are also used in several countries as an instrument to reduce risk. An apparent problem connected to this is how to reach all consumers.

Cadmium poses a special problem. High concentrations of cadmium can be found in, for instance, offal and shellfish. Recommendations on limited consumption of these foodstuffs is a possible instrument in reducing the risk for certain groups of the population. However, these foodstuffs only marginally contribute to the daily dietary intake of cadmium for most people. Generally, staple foods like cereals, grains and potatoes are the dominating cadmium sources. In these cases, recommendations to limit consumption are impossible. It should also be noticed that an already small increase in cadmium concentrations in these types of food will lead to a significant increase in the daily cadmium intake.

The options of the National Food Administration in taking steps towards decreasing the dietary cadmium intake in the general population are limited to introducing limit values for cadmium in food. However, there are some weaknesses. First, the effectiveness of such a limit value is dependent on the possiblities to actually control the levels in food. Secondly, this measure acts far from the actual source of the problem and therefore serves as an expression from the authority that a continously increasing exposure is not accepted.

The mean cadmium content in wheat in Sweden is approximately 25-50 µg/kg (Öborn, 1995). A maximum level of 100 µg cadmium/kg in cereals and cereal products is under discussion in the Council of Europe, the EU, including Sweden, and FAO/WHO. The effect of introducing such a maximum level in Sweden might be that an increased cadmium content in wheat is to some degree accepted. This is however not the case. The maximum level discussed is rather an instrument making it possible to sort out the crops with extremely high cadmium contents. In Sweden approximately 5-10 per cent of the wheat production will be excluded from the market because of high cadmium contents (Andersson, 1995). It is noteworthy that the milling industry in Sweden has taken responsibility by voluntarily introducing the maximum level for wheat mentioned above in their quality programme.

Conclusions

There are indications that the PTWI for cadmium does not adequately protect against health effects. Therefore, a continous accumulation of cadmium in the soil cannot be accepted for public health reasons. In order to minimise the risk of cadmium-induced health effects in the general population, the National Food Administration strongly proposes that necessary steps, close to the sources, are taken to prevent the increase of cadmium in the soil.

In spite of the seemingly slow increase in cadmium contamination of the soil, it is urgent to start the work with reducing cadmium emissions now. Otherwise coming generations will face the problem of how to decrease cadmium content in soil and/or crops. To refrain from taking action now might have enormous financial effects in the future.

References

Andersson A, University of Agricultural Science, Uppsala, personal communication,1995.

Buchet JP, Lauwreys R, Roels H. et al.. Lancet 1990; 336:699-702.

Ewers U, Brockhaus A, Dolgner R, Freier I, Jermann E, Bernard A. et al.. Environmental exposure to cadmium and renal function of elderly women living in cadmium-polluted areas of the Federal Republic of Germany. Int Arch Occup Environ Health 1985; 55:217-39.

Gerhardsson L, Oskarsson A, Skerfving S. Acid precipitation - effects on trace elements and human health. Sci Tot Env 1994; 153:237-45.

IPCS Environmental Health Criteria 134. Cadmium. WHO, Geneva,1992.

Järup L, Elinder C-G. Dose-response relations between urinary cadmium and tubular proteinuria in cadmium exposed workers. Am J Ind Med 1994; 26:759-69.

JECFA, The 41st Meeting of the Joint FAO/WHO Expert Committee on Food Additives and Food Contaminants. Evaluation of certain food additives and contaminants. WHO Technical Report Series No 837, 1993.

OECD, Risk Reduction Monograph No. 5: Cadmium. Background and national experience with reducing risk. Environment Directorate, Organisation for Economic Co-operation and Development, Paris, 1994.

Öborn I. Cadmium in the crops - increasing concentrations? In: Cadmium - health risks. Kemi Report 10/95. The Swedish National Chemicals Inspectorate.

Roels H, Bernard AM, Cárdenas A, et al.. Markers of renal changes induced by industrial pollutants. Br J Ind Med 1993; 50:37-48.

Staessen JA, Lauwreys RR, Ide HA, Vyncke G, Amery A. Renal function and historical environmental cadmium pollution from zinc smelters. Lancet 1994; 343:1523-7.

Strehlow CD, Barltorp D. The Shipham Report. Health studies. Sci Tot Env 1988; 75:101-34.

Fertilizer Input of Cadmium into Canadian Prairie Soils

T.L. Roberts

Potash and Phosphate Institute of Canada

Although the need for phosphorous fertilizer has been recognized in Canada since the early 1920s, phosphate (P_2O_5) fertilization did not become common until the 1950s. Usage dramatically increased from the 1960s to the 1980s, peaking at 725,000 tonnes in 1985 (Figure 1). Approximately 640,000 tonnes of P_2O_5 was sold in Canada in 1994. About 70 per cent of Canadian consumption is in the prairie provinces.

Figure 1

Phosphorus consumption in Canada (1966-1994)

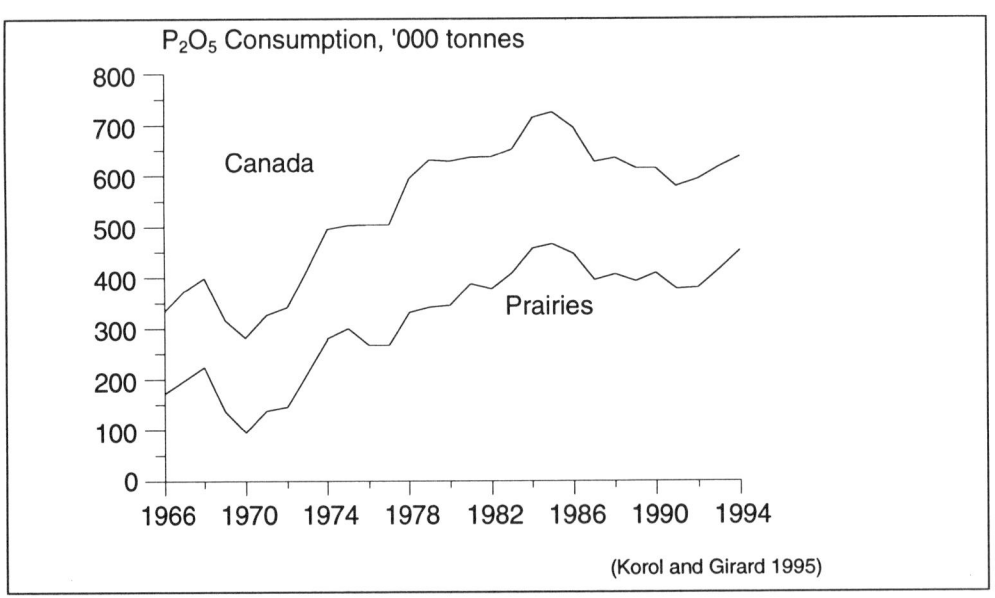

(Korol and Girard 1995)

Soils in the Canadian prairies are naturally low in phosphorous. Approximately 70 per cent of Alberta soils, 86 per cent of Saskatchewan soils, and 77 per cent of Manitoba soils test medium or less in plant available phosphorous and require fertilization (PPI 1994). And, more phosphorous is exported from these soils in crops than is returned through fertilization. The P_2O_5 shortfall (removed in grain minus applied as fertilizer) for the five-year period ending 1989 was 86,400 tonnes per year (Rennie et al. 1993). Because there is a negative phosphorous balance, phosphorous fertilizer consumption in the Canadian prairies is expected to continue at present rates or increase in the future.

About 80 per cent of the phosphorous fertilizer sold in Canada is produced from Togo phosphate rock. The balance is derived from US phosphate rock (70 per cent from Florida and North Carolina and 30 per cent from western states). Cadmium concentration in the rock varies widely depending on the source. Togo rock ranges from 42 to 80 ppm cadmium (average 55 ppm), Florida rock from 3 to 20 ppm cadmium (average 8), North Carolina rock from 20-50 ppm cadmium (average 40), and western US rock from 40 to 150 ppm cadmium (average 90 ppm).

Cadmium transfer from phosphate rock to the fertilizer varies with rock source and processing. For wet-process phosphoric acid, about one-third of the cadmium goes to the gypsum and two-thirds to the filter-grade acid (Wakefield 1980). Phosphoric acid is used to make ammonium phosphates and triple super phosphate. In the case of single and triple superphosphate, all the cadmium in the feed rock and that in the acid would remain with the finished product. Most of the phosphorous fertilizer used in western Canada is in the monoammonium phosphate (MAP) form.

Fertilizer input of cadmium to Canadian prairie soils during the past 30 years was estimated assuming: (i) MAP was the major source of phosphorous fertilizer, (ii) cadmium concentration of the MAP was 55 ppm, and (iii) the fertilizer phosphorous used was applied on the 27.5 million hectares of cropped land in the prairies (Statistics Canada 1992). For example, in 1994, 452,598 tonnes of P_2O_5 was applied over 27.5 million hectares of cropped land; that averages about 17 kg/ha of P_2O_5 or 33 kg/ha of MAP (12-51-0). The cadmium added to the soil can be calculated as:

$$33 \ kg \ MAP / ha \ X \ 55 \ mg \ Cd / kg \ MAP = 1,815 \ mg \ Cd / ha \quad (1.8 \ g \ Cd / ha) \quad (1)$$

The above estimate of 1.8 g cadmium/hectare is very general, based on the stated assumptions. Absolute cadmium additions to soil from phosphorous fertilizer in any given year could be higher or lower, depending on the actual cadmium concentration of the applied phosphate and the cropped acreage over which it is applied.

Assuming the plow layer of the soil weighs 2.6 million kg/ha (i.e. bulk density of 1.3 g/cm3 and 0-20 cm depth), a cadmium addition of 1.8 g/ha increases the cadmium concentration of the plow layer by:

$$\frac{1,815 \ mg \ Cd / ha}{2.6 \ X \ 10^6 \ kg \ soil / ha} = 6.98 \ X \ 10^{-4} \ mg \ Cd / kg \quad (0.7 \ ppb \ Cd) \quad (2)$$

In 1994, phosphorous fertilization increased the cadmium concentration of the surface soil by about 0.7 ppb. Since 1960, phosphorous fertilization has added an estimated 38 g/ha of cadmium to prairie soils. This has increased the cadmium concentration of the upper 20 cm of soil by about 15 ppb (Figure 2), or about 5 per cent above background levels. Median background levels of cadmium in the Canadian prairies have been reported at 0.3 ppm or 0.78 kg/ha (Garrett 1994).

Although cadmium is accumulating in our soils due to phosphorous fertilization, relative to background levels, the build-up of cadmium appears insignificant. Annual additions of cadmium from phosphorous fertilizers would be difficult to detect. An increase in soil cadmium concentration of 0.7 ppb is about 15 times less than analytical detection limits (detection limit for cadmium is 10 ppb, Holmgren et al. 1993).

Figure 2

Estimated cadmium input to prairie soils from phosphorous fertilizers (1960-1994)

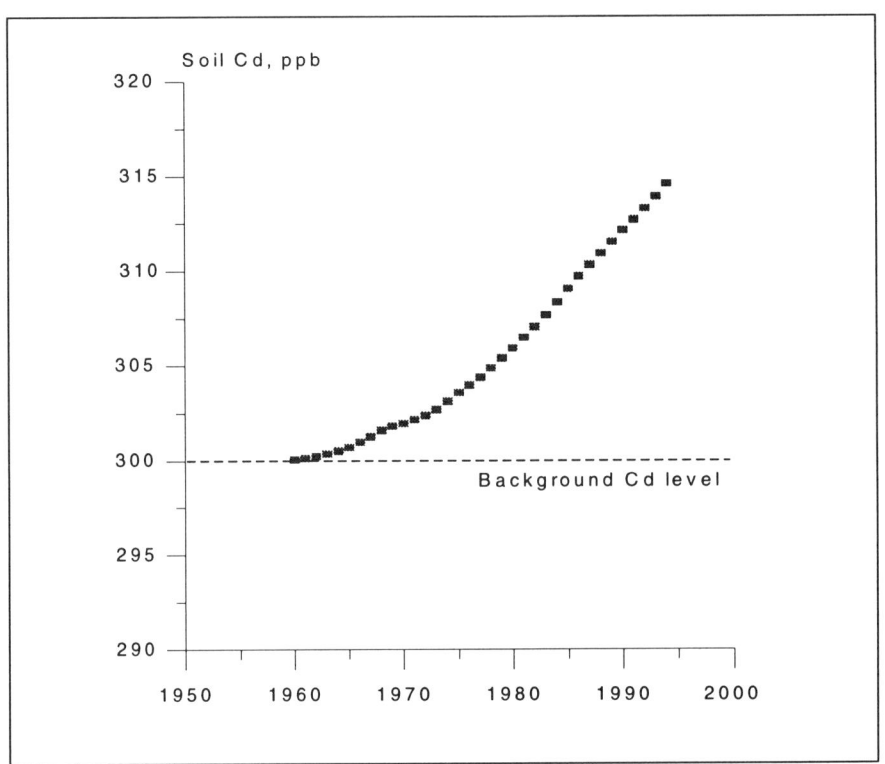

If future phosphorous fertilization continues at 1994 application rates (i.e. 1.8 g cadmium/ha/yr or 0.7 ppb), it would take about 435 years to double the average background level of cadmium that occurs naturally in our prairie soils. Phosphorus fertilization is increasing the total cadmium content of Canadian soils, but the impact or significance of the small contributions phosphorous fertilizer makes to soil cadmium is largely unknown.

Increases in total soil cadmium do not mean biological availability is increasing at the same rate. Plant uptake of cadmium is related to cadmium concentration in the soil solution rather than total cadmium concentration in the soil. Cadmium concentration in the soil solution is influenced by many soil factors in addition to total soil cadmium content. Soil pH, clay content, ionic interactions, organic matter, temperature, moisture content, compaction, aeration and the presence of other nutrients all affect cadmium availability. Plant uptake of cadmium is also influenced by agronomic practices such as tillage and crop rotation. And, different crop species and cultivars within species accumulate cadmium to different degrees when grown on the same soil.

In summary, phosphorous fertilization is a minor contributor to soil cadmium in the Canadian prairies. At current rates of application it would take several centuries for cadmium additions from phosphorous fertilization to equal background levels, which are naturally low. In our highly buffered, high pH prairie soils, cadmium uptake and accumulation by crops should not be enhanced by current phosphorous fertilization practices.

References

Garrett, R.G. 1994. The distribution of cadmium in A horizon soils in the prairies of Canada and adjoining United States: in Current Research 1994-B; Geological Survey of Canada, p. 73-82.

Holmgren, G.G.S., M.W. Meyer, R.L. Chaney and R.B. Daniels. 1993. Cadmium, lead, zinc, copper, and nickel in agricultural soils of the United States of America. J. Environ. Qual. 22:335-348.

Korol, M. and L. Girard. 1995. Canadian fertilizer consumption, shipments and trade 1993/94. Farm Income Policy and Programs Directorate, Agriculture and Agri-Food Canada.

PPI 1994. Soil test summaries: phosphorus, potassium and pH. Potash and Phosphate Institute, Better Crops With Plant Food 78(2):14-17.

Statistics Canada. 1992. Census overview of Canadian Agriculture: 1971-1991. Statistics Canada, Agriculture Division. Ottawa. Catalogue 93-348.

Rennie, D.A., C.A. Campbell and T.L. Roberts. 1993. Impact of macronutrients on crop responses and environmental sustainability on the Canadian Prairies. Canadian Society of Soil Science, Ottawa, Ontario.

Wakefield, Z.T. 1980. Distribution of cadmium and selected heavy metals in phosphate fertilizer processing. National Fertilizer Development Center, Muscle Shoals, Alabama. Bulletin Y-159.

The Role of Food Regulations in Minimising Exposure to Cadmium

Terry Spencer

National Food Authority
Canberra, Australia

Background

It has been recognised that for the vast majority of the human population the major form of exposure to cadmium occurs through food. This is true even for heavy smokers who are exposed to moderately high levels of cadmium in tobacco smoke. The only exceptions may occur in certain industrial environments: for example, workers in zinc smelters.

Exposure by the latter route can, and has been, minimised by implementation of effective occupational health and safety procedures in the form of barrier controls, education programs and containment processes. These procedures have been enforced in many countries through the implementation of air quality standards. Similarly, many countries have established controls to minimise the cadmium content of materials such as phosphate-based fertilizers, animal manure, sewerage sludge, plastics, coatings and animal feed.

In most OECD countries, one notable omission from the range of regulations and standards to control cadmium is a set of standards for the maximum permitted concentrations (MPCs) of cadmium in food for human consumption. This situation seems anomalous in view of the important role that food plays in determining the level of exposure to cadmium.

The Australian position is that standards for cadmium in food play a major role in minimising exposure to this heavy metal. This has been further recognised through our recent proposal to develop a cadmium minimisation strategy in which the review of food standards is a vital element. A discussion of this and other elements of this strategy can be found in a companion paper entitled "Developing an Australian Cadmium Minimisation Strategy" (see Session C).

The present paper will provide an overview of food standards for cadmium, including what they mean, how they are set from a regulatory and public health point of view, and the status at an international and national (Australian) level. It will also describe the way in which standards can be used to justify, monitor and evaluate the imposition, maintenance and severity of any minimisation strategies.

Cadmium and health

It is important to note that cadmium food standards, by themselves, are not a primary health standard. The primary health standard for cadmium is the Provisional Tolerable

Weekly Intake or PTWI, based on total exposure to cadmium from all sources. The PTWI for cadmium has been determined by the FAO/WHO Joint Expert Committee on Food Additives (JECFA) to be 7 µg per kilogram bodyweight per week. Although cases have been put forward for increasing or decreasing this figure, it is generally accepted internationally.

Contaminants vs residues vs pollutants

Cadmium occurs naturally in the environment and is an unintentional contaminant of food. The Codex Alimentarius Commission (Codex) definition of a contaminant is "any substance not intentionally added to food, which is present in such food as the result of production, manufacture, processing, preparation, treatment, packing, transport or holding of such food or as a result of environmental contamination." Obviously cadmium, whether it be from industrial pollution, sewerage sludge, phosphatic fertilizers or other sources, fits into this definition.

In food law, the term residue is used in relation to agricultural and veterinary chemicals - there is an element of intentional addition to the raw food commodity, e.g. by spraying or injection, during production such that the residue is an integral part of the final food.

The term pollutant in the context of food and food hygiene is usually reserved for material that is present as a result of industrial activity and overtly affects the product. Examples that come to mind include filth (there are also standard definitions for filth) and the presence of oils and other visually obvious materials.

Setting MPCs

The maximum permitted concentrations (MPCs) for a contaminant, such as cadmium, are determined by reference to two criteria:

- The level should not result in adverse health effects; and

- The level should encompass the usual range of cadmium levels in the commodity.

The corollaries of these two criteria are:

- There needs to be a means by which the total dietary exposure to cadmium can be determined. The latter process is called dietary modelling (see below); and

- Data is required on levels of cadmium that are found in the range of foods that form the diet (usually on a commodity basis).

Calculated dietary uptake of cadmium

The dietary intake of cadmium for both average and extreme consumers can be estimated using data available from surveys of the composition of the average diet and cadmium levels determined in the components of the diet. Such calculations form the basis of market basket (or total diet) surveys that are carried out by many countries, not only for cadmium but also for a variety of contaminants and residues. Intake data calculated from such surveys tends to over-estimate the actual intake by between 50 and 100†per cent, due to the conservative assumptions used in the data evaluation.

Comparative dietary uptake

The results from the above surveys are collated under joint sponsorship of the United Nations Environment Programme, the Food and Agriculture Organization, and the World Health Organization in cooperation with the Global Environment Monitoring System (GEMS). Although the results available from these surveys suffer from a degree of uncertainty, mainly associated with assumptions regarding dietary intake, they can at least be used as a semi-relative means to gauge exposure to cadmium on a historical and country-to-country basis.

Dietary modelling

The National Food Authority, as well as a number of other food regulatory bodies, are actively developing more rigorous methods, often referred to as dietary modelling, for calculating the dietary intake of contaminants, pesticides and veterinary drugs plus all types of food additives.

International food regulation

Codex, a joint initiative of the Food and Agriculture Organization of the United Nations and the World Health Organization, develops de facto international standards that facilitate international trade while protecting the health of consumers. Since the completion of the Uruguay round of GATT talks and the establishment of the World Trade Organization (WTO), the standards developed under the umbrella of Codex have assumed greater importance in the context of international trade.

International regulation of cadmium MPCs

As indicated above, few countries have developed and adopted food standards for cadmium. The main reason for this is that many countries are guided by Codex standards and Codex has yet to ratify any standards for cadmium.

Codex, through two of its subsidiary bodies, the Codex Committee on Cereals, Pulses and Legumes (CCCPL) and the Codex Committee on Food Additives and Contaminants (CCFAC), has been dealing with the issue of cadmium standards for some years.

At one stage of its process, the CCCPL member countries agreed, after much debate, to adopt a "provisional guideline level" of 0.1 mg/kg. This decision was reversed at the 1994 meeting of the CCCPL with a recommendation that a sub-committee re-assess the suggested level.

The CCFAC, at its meeting in March 1995, discussed a position paper on cadmium prepared by France.

Among other things, this paper made the following points:

- On commercial (trade) grounds ..... "International efforts to reduce exposure to cadmium and standardize cadmium limits in foods are entirely justified."

- "The CCFAC should propose international limits for cadmium in foods ...."

- "Any reduction in food cadmium content requires the implementation of a series of measures mostly geared towards preventing the release of cadmium in the environment."

These points are consistent with the approach that has been taken in Australia, namely steps to develop a strategy for minimising the exposure to cadmium that encompasses all elements that are affected.

CCFAC also has been progressing the Codex General Standard for Contaminants (GSC). Codex member countries have agreed that priority for inclusion in the GSC will be based upon evidence of a risk to public safety from the contamination of food commodities in international trade. Cadmium was discussed in the context of this Standard and an in-principle agreement reached to progress, albeit slowly, Codex standards for cadmium in food (see above).

Australian food standards for cadmium

Before 1980, limits for cadmium in food were controlled by a general entry in the Australia Food Standards Code under "metals not specified", those being 5.5 mg/kg in solid foods and 0.15 mg/kg in beverages. Since the establishment of the first set of specific standards for cadmium in food in 1980, by the predecessor of the National Food Authority, the National Health and Medical Research Council, there have been a number of variations and additions resulting in the values shown in the Table 1.

These standards are a mixture of commodity items, e.g. fish and processed or semi-processed foods, e.g. beverages and bran. Many of the staple foods (cereals, vegetables) fall by default into the general category of "foods not containing a food otherwise specified". The latter entry is included, instead of an "all other foods" entry, to allow calculation of "adjusted" MPCs, based on proportions, for processed multi-component foods.

Table 1

Australian food standards for cadmium

Food	MPC (mg/kg)
Beverages and other liquid foods	0.05
Bran	0.2
Cocoa	0.5
Cocoa paste	0.35
Chocolate	0.25
Drinking chocolate, powder	0.25
Crustaceans	0.2
Fish	0.2
Edible offal other than liver	2.5
Liver	1.25
Meat muscle	0.2
Molluscs	2.0
Seaweed (edible kelp)	0.2
Water	0.005
Wheat germ	0.2
Foods not containing a food otherwise specified	0.05

Review of cadmium MPCs in Australia

The National Food Authority, the body responsible for formulating food standards in Australia, is currently undertaking a review of MPCs of cadmium in all foods. The main reason for the review is that a number of surveys have indicated that not all foods available within Australia can adhere to the current MPCs. This is especially the case for foods in the "foods not containing a food otherwise specified" category.

These data suggest that the current MPCs for cadmium do not represent actual or achievable levels, and that the standards should be amended to reflect the lowest reasonable levels of cadmium achievable that are consistent with public health and safety. As part of the review, consideration is being given to classifying foods into groups and sub-groups that have similar characteristics and potential for cadmium content.

The review serves a second purpose, in that it has focused attention on the need to minimise cadmium in the Australian diet and the consequent decision to develop a strategy.

Cadmium review strategy

Consistent with the current international approach to dealing with contaminants and setting permitted levels, a strong emphasis will be placed on risk assessment and risk management. The elements of this approach are consistent with those of Codex, GATT/WTO agreements and will rely heavily on dietary modelling protocols to determine those components that contribute the greatest amounts to the dietary intake of cadmium. For example, there is a case for only setting standards for those items that contribute a significant proportion of the total dietary intake of cadmium. This could obviate the need to set MPCs for minor food items, e.g. exotic plant materials, and the need to create an "all other foods" category.

Reducing exposure to cadmium through the use of standards

Food standards for cadmium are used in a direct manner, under some circumstances, to reduce exposure to cadmium in food. Current examples include:

- the European Union requirement concerning the maximum cadmium level in animal offal used for human food. Within Australia, this has resulted in significant research on factors that affect the cadmium content of animal offal and a subsequent prohibition on the use and export of offal derived from animals above a certain age;

- the requirement by the Peanut Company of Australia, the major marketing company for peanuts within Australia, that their growers participate in a cadmium monitoring program and undertake test plantings in new growing areas before full-scale production to ensure compliance with the Australian standard; and

- the requirement by the Australian crisp and french fry processors that only potatoes that meet the current MPC are acceptable for use in their products. This has resulted in growers changing agronomic practices, e.g. use of non-saline irrigation water, to achieve the desired low cadmium levels.

In all these cases, there is a minor economic penalty that is initially borne by the producer and subsequently passed on to the consumer in some form.

Cadmium, risk reduction and minimisation strategies

Australia believes that, while assessment of the intrinsic hazard of a substance can be conducted at an international level, risk assessment based on specific exposure measures and subsequent risk management should occur at a national or regional level.

Suitable measurable parameters are required to undertake meaningful risk assessment. In the case of cadmium, the only obvious candidates are associated with cadmium levels in food. They include the actual cadmium levels in the range of foods that are consumed combined with dietary intake patterns, i.e. dietary modelling. Other parameters are also important in calculating the final level of exposure - these include area-to-area differences in the cadmium content of food and variations in dietary intake patterns.

The latter information, particularly that relating to area-to-area (or region-to-region or country-to-country) differences, is important in undertaking risk management. In simple terms, it should not be necessary to impose universal, immediate and extreme restraints regarding cadmium use or content. This is especially so when the available evidence, in the form of cadmium levels in food, shows that, on an area-to-area, region-to-region or country-to-country basis, this may not be required (on health grounds) or justified (on economic grounds).

The role of radmium food standards

So where do food standards for cadmium fit into such a scenario?

1. They should provide the trigger for implementing controls on environmental cadmium sources in an area, region or country.

2. They should be the means by which the effectiveness of any minimisation strategy is judged, i.e. usual distribution range compared to the MPC.

3. They should determine whether food with particular levels of cadmium should be traded and consumed.

For these reasons, Australia believes food standards have a central role to play in limiting the exposure to cadmium, given that the major source of exposure comes from the consumption of food.

REPORT OF SESSION C

ACCUMULATION IN AGRICULTURAL SOILS AND CONTENT IN FOOD AND HUMAN INTAKE

The papers presented in this section were given during Session C.

Initial considerations

Whilst interest in cadmium is centered upon potential health effects through exposures in food and smoking, it was outside the scope and expertise of this workshop to discuss the nature of the actual or potential adverse effects on human health and ecosystems.

The group also considered the topics in question in the context of the outcomes of the sources workshops, especially (E) on mineral fertilizers and other sources, to which reference is made in this paper.

ACCUMULATION IN AGRICULTURAL SOILS

The problem

The group agrees that there is a need to minimise the accumulation of cadmium in soil, which can be achieved by a number of means including the reduction of anthropogenic cadmium inputs.

The need for and extent of the reduction of anthropogenic inputs will vary from country to country, depending on factors such as current rates of cadmium accumulation in soils.

Potential/actual measures for reduction

Standardization and implementation of techniques to monitor rates of accumulation of cadmium in soils:

- understand landscape variability
- establish monitoring sites
- determine the dynamics governing input/output relationships

Assess options for reducing cadmium in phosphatic fertilizers - and other soil amendments:

- regulation
- voluntary agreements
- economic instruments
- decadmiation technologies
- sources of low cadmium phosphate
- labelling of product
- application rates vs. levels for adequate phosphorous
- education

Define vulnerable areas and apply appropriate agricultural practices.

Implications and cost/benefit

Minimize accumulation of cadmium in soils.

Cost of monitoring programme.

Potential economic stimulus for farmers to consider cadmium levels in products.

Impacts on farmers' costs of production.

Impact on production and trade in fertilizers.

Cost of labelling and informing farmers on cadmium levels in fertilizers.

Compliance cost.

Response

Possible strategic measures include:

- assess cadmium accluation rates;
- reduce cadmium in fertilizers and other soil amendments;
- define vulnerable areas.

Possible strategic implications are:

- impact on trade;
- impact on fertilizer producers;
- social benefits/costs.

CONTENT IN FOOD/ INTAKE INTO HUMANS

The problem

Recognise the need in the long term to minimise intake of cadmium from diets and smoking.

It is uncertain whether cadmium levels in food are increasing because

- there is insufficient time to establish trends;
- trend data are not always reliable and are insufficient for analysis and interpretation;
- there is difficulty in comparing data.

Average dietary cadmium intake appears to be within current PTWI.

Potential/actual measures for reduction

International standardisation of techniques for assessing and presenting cadmium in foods and diets: various Codex Committees are in the process of doing this.

Countries may consider establishing standards for major food staples.

Need for open transfer of information on cadmium in foods and diets internationally.

Need for education: producers and community at both national and international levels.

Implications and cost/benefit

Recognise existing and potential trade implications, e.g. "eco-dumping"

Recognise existing and potential food industry implications

Recognise commercial implications of information availability internationally

Potential increased consumer cost

Response

Possible strategic measures include:

- consider population groups at risk;
- develop local, regional and national approaches, as appropriate;
- develop food standards;
- encourage free international exchange of information

SESSION D

UPTAKE INTO CROPS AND BIOAVAILABILITY

Cadmium Accumulation and Availability in Agricultural Land and the Effects of Land Use Changes

Pierre del Castilho, Jan Bril, Paul Römkens and Oene Oenema

DLO Research Institute for Agrobiology and Soil Fertility
The Netherlands

Abstract

In the present overview the determining factors for the soil solution cadmium concentration are discussed. Soil characteristics have a strong influence on the uptake and leaching of cadmium. The activity and concentration of cadmium in the soil solution may vary, depending on the input and the specific physico-chemical soil factors. The soil solution composition and the speciation of cadmium determine the uptake by plants and the extent of leaching. Therefore, these factors also have an impact on cadmium retention and accumulation.

To explain the cadmium uptake by leafy crops a survey with 50 Dutch agricultural soils was done. Strong relations between the soil pH, the concentration ratio of $Cd_{leafy\ crops}/Cd_{soil}$, and the calculated ratio of $Cd_{soil\ solution}/Cd_{soil}$ were found. Temporal variation of soil solution cadmium concentrations was determined in 3 fallow agricultural acidic loamy sand soils with similar soil cadmium levels (around 0.1 mg cadmium kg^{-1}) on seven bi-monthly occasions. In the topsoil - rich in organic matter - the soil solution pH was the main factor for the cadmium concentration in the soil solution. In the subsoil (40-80 cm depth) also dissolved organic carbon (DOC) correlated positively with cadmium soil solution concentrations.

Spatial variation was determined in a recent Dutch national survey covering arable land, woodland and nature areas. Soil solution pH values of nature areas were up to 3 units lower and soil solution cadmium concentrations up to 10 times higher, compared to agricultural areas. A chronosequence study (seven woodlands, created between 1 and 70 years ago on former agricultural land) showed a decrease in soil pH by 2 units within decades, and an increase in DOC by a factor 3.

A model was used to predict the effects of land use change. In this theoretical approach the atmospheric deposition of cadmium was the only input source over 100 years. The calculation was done for a sandy soil with, at the start of the calculation, a soil solution pH value of 6. An 8-cm litter layer was allowed to build up in about 15 years. The variation of the cadmium content in the soil profile, the yearly plant uptake and the soil solution cadmium concentrations were calculated. Because of the drop in pH and the loss of cation exchange buffering capacity, there was a substantial movement of cadmium through the profile within a 50-year period. The model will be extended to include more parameters.

It is concluded that soil pH is one of the important factors determining short-term and long-term variations in uptake and mobility of cadmium. Other soil physico-chemical and biological characteristics must be considered as well for a precise prediction of the activity, the concentration, and the speciation of cadmium.

Introduction

Cadmium accumulates in agricultural soils because it strongly sorbs to soil and the input by atmospheric deposition and fertilizers is larger than the output via crop removal, erosion and leaching. Input estimates of cadmium on agricultural soils by mineral fertilizers in the Netherlands range from 6-25 and by atmospheric deposition from 2-3 g ha^{-1} y^{-1}; output via crop removal and drainage is about 3 g ha^{-1} y^{-1}. To prevent toxic levels in the long term, cadmium content regulations for fertilizers and organic wastes are necessary.

When agricultural land is taken out of production, atmospheric deposition becomes the only cadmium source. Langeweg (1989) estimates the atmospheric deposition to be 2.8 g ha^{-1} y^{-1}. Without plowing, like in nature areas, a high accumulation is expected in the thin top layer. However, when land use is changed, a litter layer will build up, which dilutes the level, while accelerated leaching may also decrease the levels. The combined effects of these processes will be discussed using a modeling approach.

First this paper gives a short overview of factors that contribute to cadmium accumulation and availability in soil. Second, it presents some recent findings about the effects of changing agricultural land into forest and nature preserves on cadmium mobility and availability.

1. Cadmium accumulation in agricultural soils in the Netherlands

Van Driel and Smilde (1990) estimated the accumulation rate of cadmium in the Netherlands by calculating the difference between the estimated annual input and output (Figure 1). The input by mineral fertilizers was approximately 6 g ha^{-1} y^{-1} when only mineral fertilizers were used. Combinations of different organic and mineral fertilizers (excluding sewage sludge) lead to a lower accumulation.

More cadmium input estimations were made by Van Erp and Meeuwissen (1994). These authors calculated an input of about 6 g ha-1 y-1 for a potato/beet/corn rotation. The estimated cadmium input for growing vegetables was about 25 g ha^{-1} y^{-1}. The accumulation of 20 g ha^{-1} y^{-1} during 100 years would result in a rise of 0.5 mg cadmium kg^{-1} soil (dry weight), assuming that cadmium remains in the 30 cm plow layer. In Dutch arable soils the cadmium levels are usually below 1 mg cadmium kg^{-1} soil.

Figure 1

Cadmium balance for arable soils: data from Van Driel and Smilde (1980). Effect of organic and inorganic fertilizers.

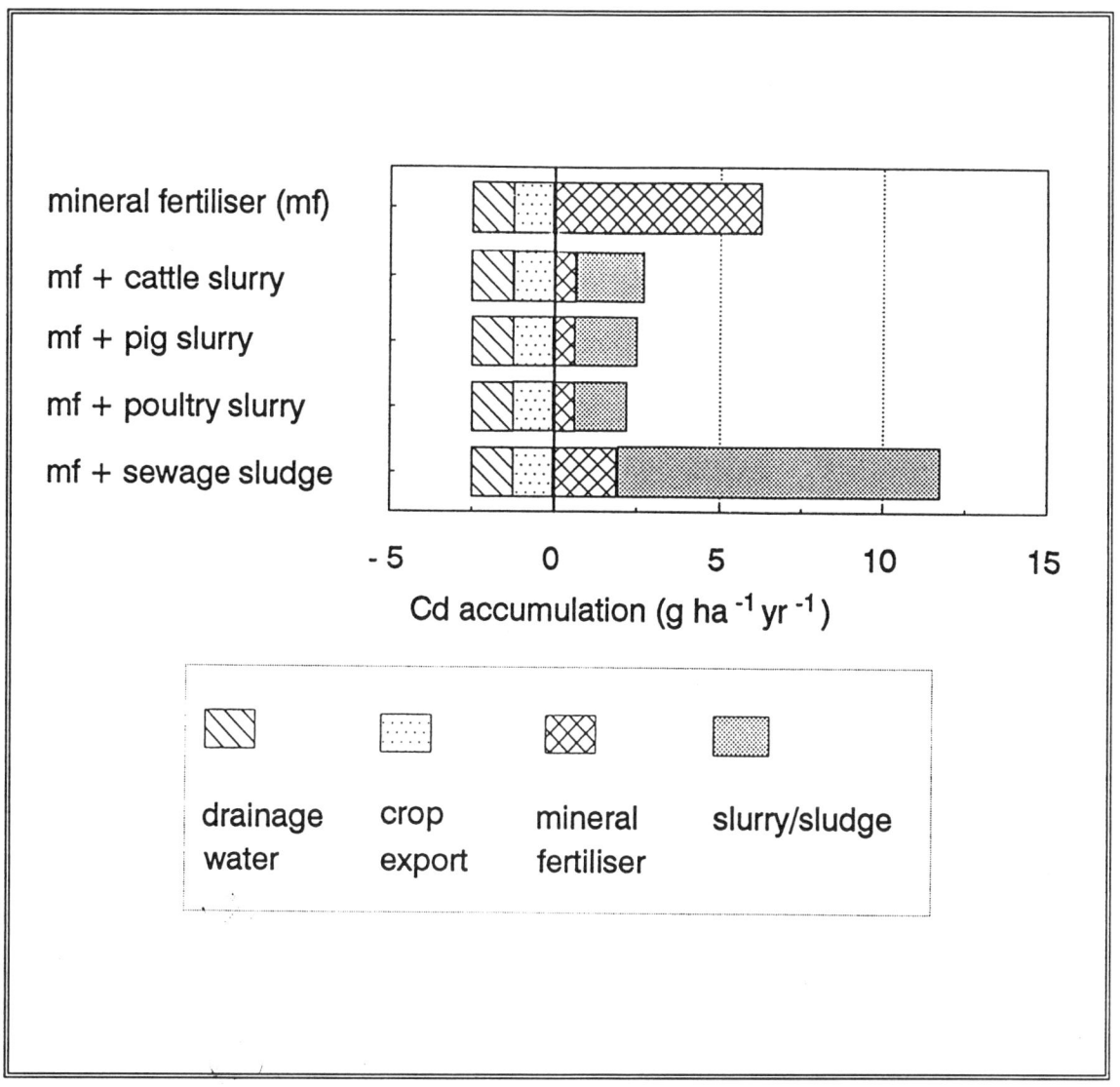

In soils with a strong cadmium binding capacity, like most agricultural soils that have pH 5-7 and are regularly limed and fertilized for optimal agricultural production, accumulation is high. The cadmium uptake by plants from soil may become important only at pH values below 5. Transfer of cadmium from soil to crop at pH values above 5, even by fast-growing accumulating species like leafy vegetables, is low (Del Castilho and Chardon, 1995). Cadmium in agricultural crops and in grass for animal feeding originates almost exclusively from soil (Dalenberg and Van Driel, 1990). In nature areas the contribution by atmospheric deposition may be important: the cadmium uptake by grass in nature areas, because of its long residence time, may be dominated by the aerial route.

In the Netherlands the soil sanitation standards as well as agricultural recommendations regard the cadmium level of a soil and the soil type. Dutch farmers are advised by the Ministry of Agriculture, Nature Management and Fisheries to determine the cadmium content of their crops in case the Signal Value (a soil cadmium level depending on the combination of soil type and use) is exceeded. Soil type and pH influence the phyto-availability of cadmium, probably by affecting the tendency of the metal to go into solution. Dissolved metals are more readily available than metals in the soil solid phase. Therefore, the agricultural recommendation ought to be based on cadmium in the soil solution, which is a much more direct approach of availability than cadmium in soil. However, information about the cadmium speciation in the soil solution may also be required to determine the activity/concentration of the most-available cadmium species.

2. *Effects of land use changes*

The pollution of nature areas by heavy metals differs in many respects from agricultural soils. In the Netherlands these soils are mainly situated on unlimed, acidic sandy soils. Current policy in the EU is to set aside a significant part of the area of agricultural soils because of the costly overproduction. Policy in the Netherlands is to create more nature areas and forest lands. Qualitative changes in cadmium accumulation are to be anticipated and are summarized in Table 1.

Table 1 indicates that the cadmium concentrations and cadmium levels in the topsoil may change drastically. However the biological and chemical changes to be anticipated with changing land use will also change the availability and mobility of cadmium. A quantitative approach including the change of ecology and soil chemistry would be needed for a more precise prediction of the effects.

Table 1

Expected initial changes as a result of turning agricultural land into rough grassland or deciduous forest

Change of land use	Effect soil Cd	Effect solution Cd
- No harvesting	-	-
- No application of manure/fertilizers	+	+
- No liming	+	-
- No plowing	-	?
- Increase of scavenging surface (no harvesting; continuous enhanced collection of air dust by grass or trees)	-	-

+ = *positive effect (decrease of soil or soil solution [Cd])*
- = *negative effect (increase of soil or soil solution [Cd])*

To illustrate the effect of forestation of arable land on soil pH and dissolved organic carbon (DOC), eight forests were investigated. All forests had been planted on sandy arable soils and varied in age from 1 to 70 years. Soil solution samples were collected from the 0 to 30 cm soil layer. The soil solution was obtained by centrifugation of bulked soil samples (Del Castilho et al., 1993a). The changes in soil pH and DOC concentration with time are shown in Figure 2a. After termination of liming, the soil pH decreased to 3.5-4 within 3 to 4 decades, indicating a rapid acidification of the topsoil due to a lack of buffering capacity of these sandy soils. Despite the lower soil pH, the DOC concentration increased with time. It has been shown by Bergkvist (1987) that under field conditions metals like copper and lead are mobilized in the presence of DOC. The increase from an initial DOC concentration of 5-10 mg C L^{-1} to 50 mg C L^{-1} within 40 years may therefore be of significance for the mobility of metals that form stable aqueous complexes with DOC. However, if the soil organic matter content would also increase, the capacity of the soil to retain metals would counter-balance this effect.

The acidification due to the land use changes will lead to a significant increase in the dissolved cadmium concentrations. To illustrate the effect of pH (and DOC) on the cadmium soil solution concentrations results of a national survey are presented (Figure 2). The soil samples were collected, and soil solution was expelled by centrifugation in winter (February-March 1991). The sites selected represent the majority of Dutch soil use/soil type combinations: grassland, crop production, fruit tree stands, and forested sites on sand, clay and peaty soils. It was found that the solubility of cadmium was controlled mostly by soil pH and, to a much lesser extent, by DOC. Despite the wide range in dissolved cadmium concentrations at a given pH, there is a significant increase with a decrease in soil pH. Especially below pH 5, dissolved cadmium concentrations increased rapidly. This is due to pH-dependent desorption from the soil surface.

Figure 2

Change of pH and DOC in the soil solution (a) under Dutch sandy soils at various periods after turning production land into forest land; and of cadmium as a function of pH (b): Data from Römkens and De Vries (1995).

The cadmium concentrations near the question mark (orchards) are close to the Intervention Value for cadmium (the Dutch sanitation standard for ground water); many concentrations (right hand side of this figure: woodlands, heather at low pH) exceed the Intervention Value.

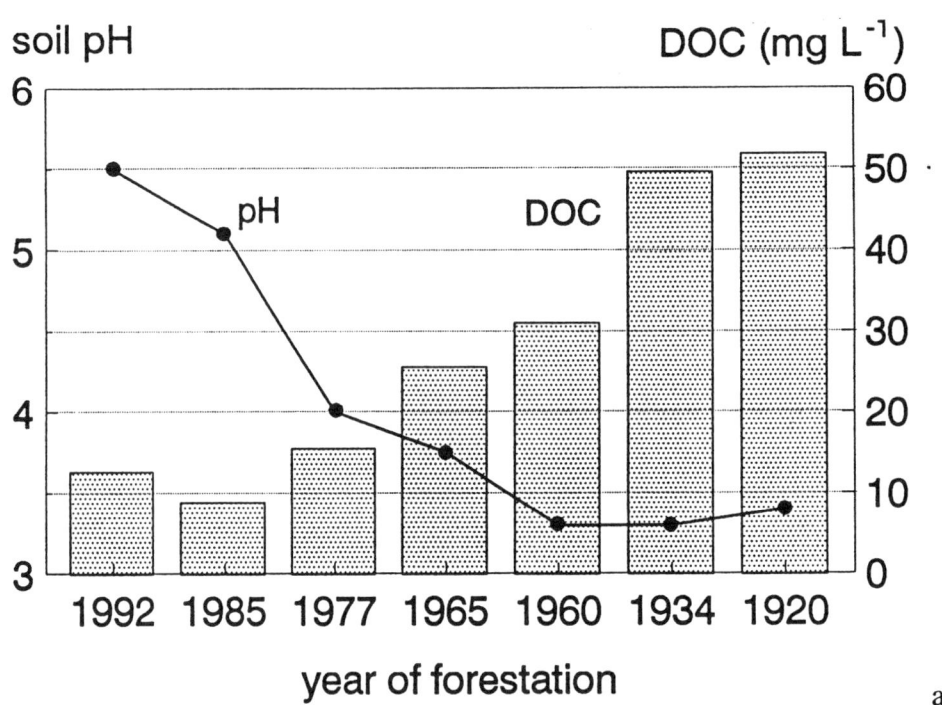

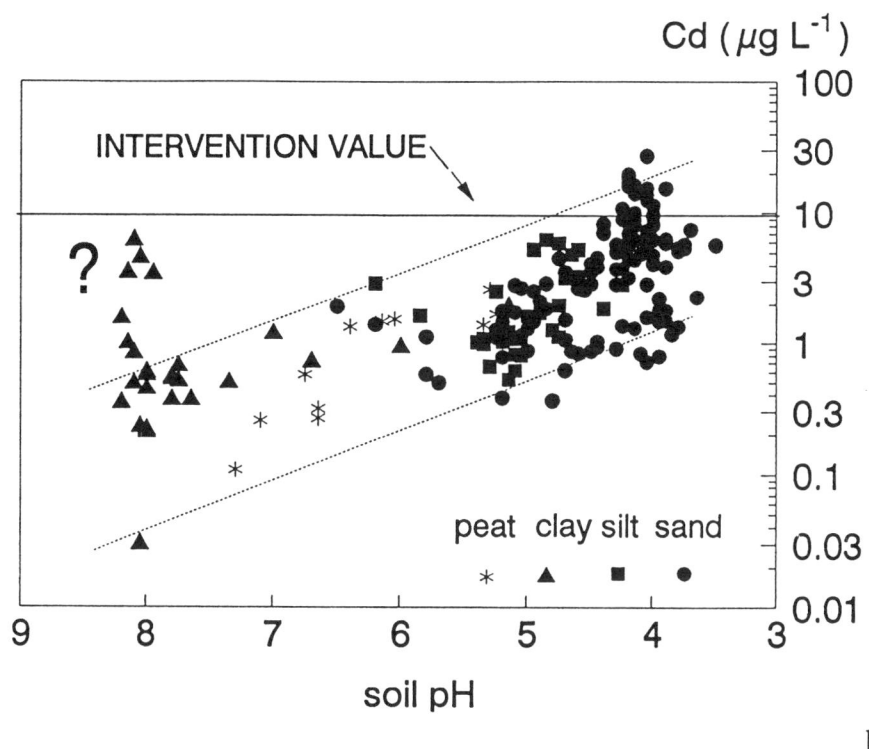

b

A study on temporal variations (Del Castilho et al., 1993a) in unlimed, fallow agricultural acidic loamy sand soils showed that the soil solution pH was the most important factor determining the cadmium concentration in the soil solution. Between 40 and 80 cm depth, DOC also had a cadmium-dissolving effect.

3. Cadmium uptake by vegetables from soils

Existing data sets on cadmium content of leafy crops (endive, lettuce and spinach) and soil data (per cent organic matter and clay, soil pH, cadmium content) were evaluated to derive the major factors controlling cadmium phyto-availability. Cadmium concentrations in the soil solution were calculated using a pH dependent Freundlich sorption model (Del Castilho and Chardon, 1995). Multiple regression calculations were done do explain the cadmium crop content or cadmium crop/soil transfer factor by: soil cadmium content, clay and organic matter content and soil pH. No direct correlation ($r = 0.02$) was found between soil and plant cadmium levels. The calculated concentration ratios Cdsolution/soil as a function of soil pH (pHKCl) and soil cadmium levels are shown in Figure 3. Calculated concentrations in the soil solution (not shown) correlated with the plant cadmium levels. Soil acidity clearly has more effect on the ratios than the soil cadmium level itself (Figure 3).

Including texture parameters (per cent clay and organic matter) did not increase the potential of the model for predicting the cadmium concentration of soil/water mixtures, although it must be realized that the model was calibrated using statistics without soil chemical (e.g. thermodynamic) considerations.

The Dutch Ministry of Agriculture, Nature Management and Fisheries advises farmers to determine the cadmium content of their products when the soil cadmium content (for their soil type/use combination) exceeds a certain Signal Value. These values are, e.g., for vegetable production on a clay or peaty soil 1.0, and on sand 0.5 mg cadmium kg^{-1} (soil, dry weight (DW)). For grassland these values are higher: 3 and 2 mg cadmium kg^{-1} (soil, DW), respectively. Although we think it is a good initiative to advise farmers in cases of suspected contamination by cadmium, we propose to use a more direct approach. This involves measurement of cadmium in the soil solution (Del Castilho and Chardon, 1995), which is assumed to be a more direct indicator of plant uptake (Figure 4). Figure 4 shows the correlation between soil pH and $\log([Cd_{Cd\ plant}]/[Cd_{soil}])$ for endive; for lettuce similar transfer factors have been found. Below pH 7.2 also for spinach similar results were found, but above pH 7.2 relatively high transfer ratios were found.

Figure 3

Calculated concentration ratios for cadmium solution/soil as a function of pH and cadmium level of the soil: data from Del Castilho and Chardon (1995).

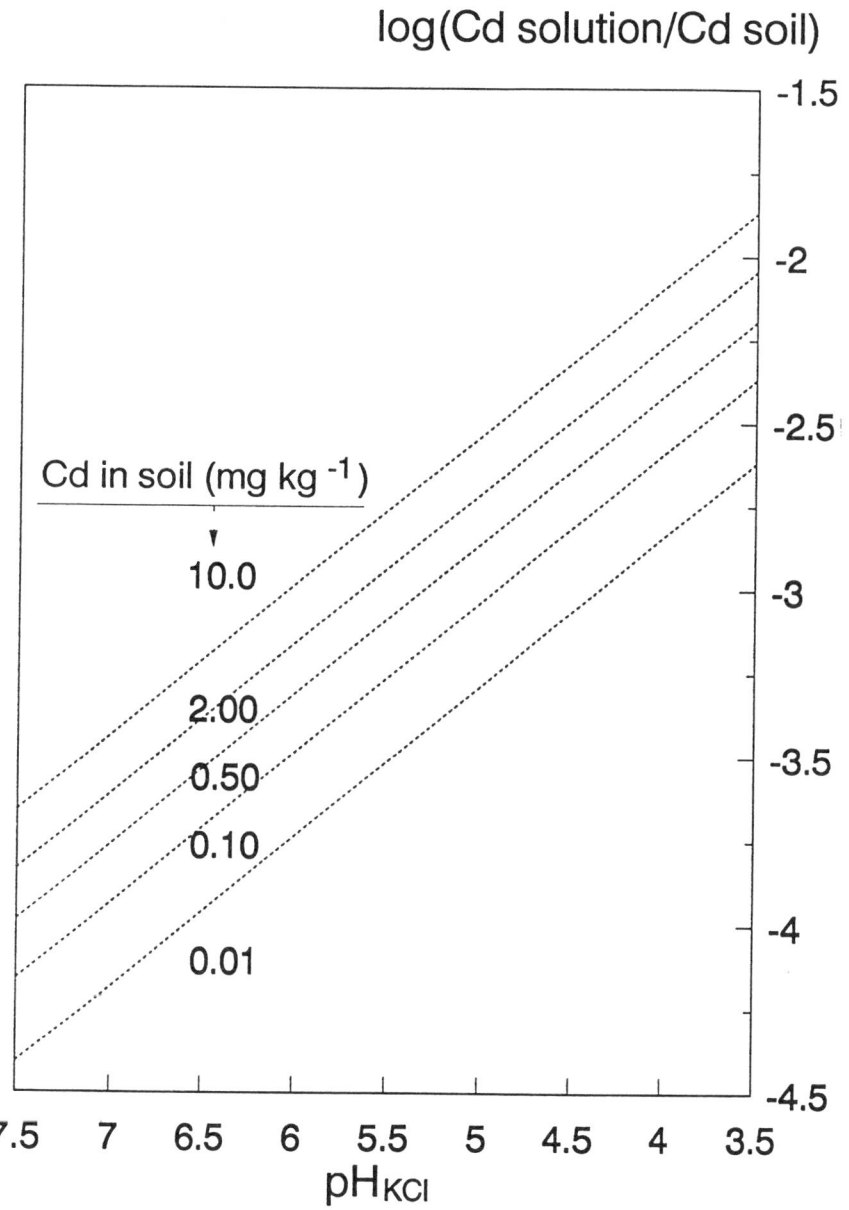

Figure 4

Cadmium transfer factor (log) for endive in a variety of soils as a function of pH: data from Del Castilho and Chardon (1995).

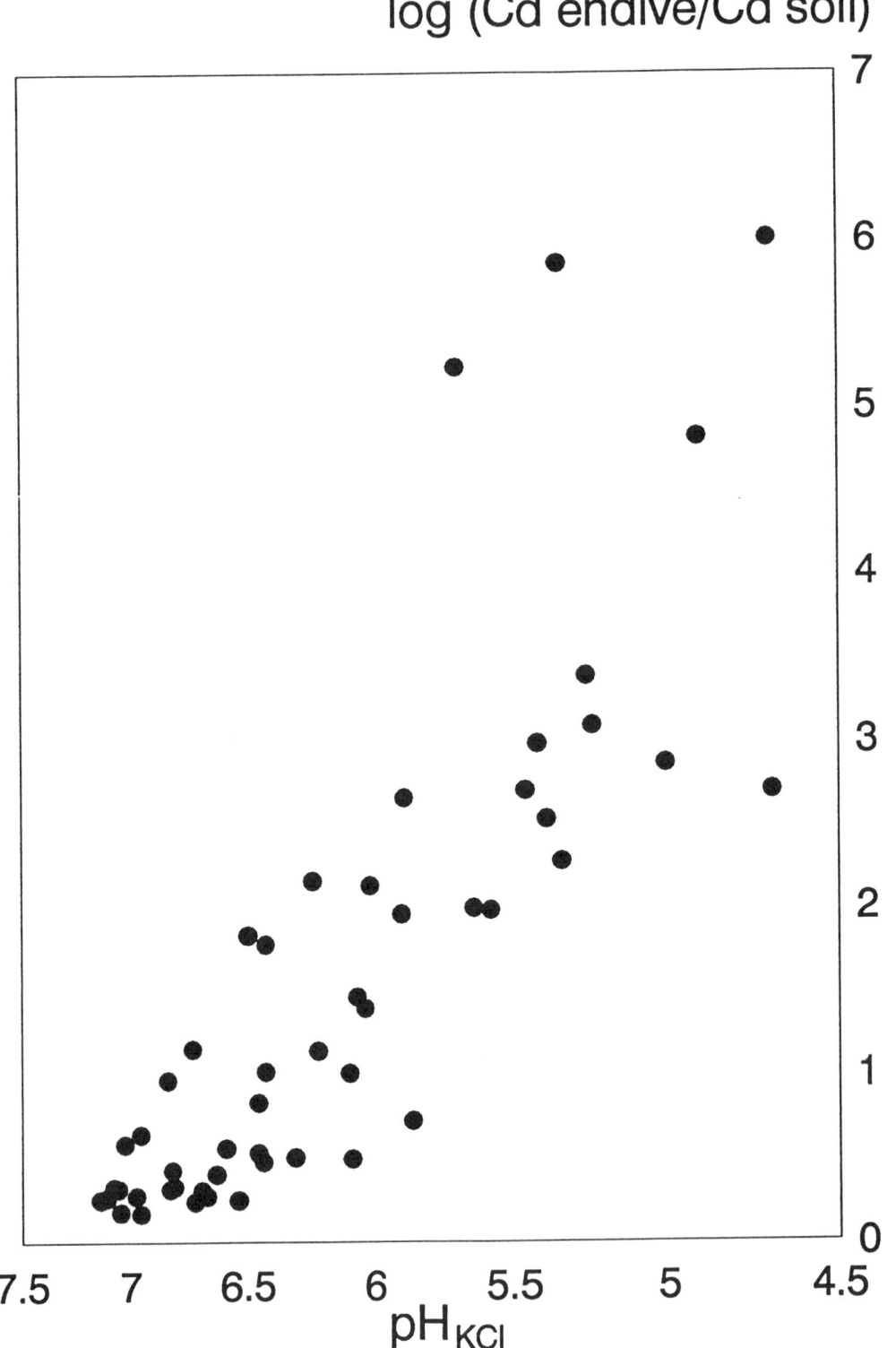

4. Cadmium speciation in the soil solution

From a theoretical point of view, larger uptakes and effects are to be anticipated, especially when the activity or concentration of the aqua Cd^{2+} species increases. It has been stated that ion pairs like $CdCl^+$ may be equally available as the free Cd^{2+} ion (Li et al., 1994). Above pH 5 most cadmium is sorbed onto the soil particle surfaces, and low concentrations in the soil solution occur. The availability for plants was low in such cases (Del Castilho and Chardon, 1995). Del Castilho et al. (1993b) expected a relatively low availability of cadmium species where cadmium is associated with molecules of high molecular weight, mainly because of their small diffusion velocity. The observed increase in DOC after land use change (Figure 2a) would decrease the bioavailability of cadmium if the DOC is (mainly) of a high molecular weight type and able to bind cadmium.

5. Model predictions of land use changes

Changing land use has several consequences which operate simultaneously and affect the soil chemistry and the soil biology of cadmium. The acidity itself and the expected increase in metal concentrations, or metal ion activities, are both likely to change the ecosystem. Evidently, the complexity of biogeochemical changes following land use change requires a comprehensive approach, i.e. a physico-chemical model to predict the cadmium concentration in the soil solution. In the following we present the definitions and results of the coupling of a simple plant uptake model to two chemical equilibrium models and a soil water transport model.

Bril (1994) used a soil chemical approach including thermodynamic equilibrium theory, together with an excess Free Energy function which accounts for surface heterogeneity as the soil chemistry module. The numerical physico-chemical model was calibrated using data from batch and column sorption experiments.

The model thus-obtained behaves like a Freundlich isotherm. The Freundlich K (K_F) is a function of CEC, organic matter content (per cent, w/w), clay content (per cent, w/w), and pH. The distribution of cadmium over the solid and liquid phase [Equation 1] depends on the value of K_F and the calcium activity (and aluminum activity in acidic soils).

$$M_{Cd} = K_F \left((a_{Cd}) / (a_{Ca})^{0.5} \right)^{0.82} \qquad \text{(Equation 1)}$$

where

M_{Cd} = cadmium concentration level in soil, expressed in mol cadmium kg^{-1} soil

aX = activity of metal in solution, expressed in mol X m^{-3}

After solving the equation using activities, the dissolved concentrations can be calculated using a speciation calculation, for which the macro-chemistry of the solution must have been analyzed (main cations and anions, including organic ligands like humic and fulvic acids).

This chemical model was coupled to a one-dimensional soil water transport module. The present calculation considers a 20-layer soil column of a total depth of 1 meter. The chemical activity of the humic and fulvic acids in the soil solution is dynamically simulated in a separate module, based on a (quasi-thermodynamic) regression model for DOC in Dutch sandy soils. To include seasonal effects of leaching and plant uptake, timesteps for the simulation were 3-month periods.

The model for cadmium plant uptake allows to choose between uptake related to the total dissolved concentration, or uptake related to the activity of cadmium species in the soil solution. In principle, uptake of magnesium and other basic cations may be included in the calculation. In the present calculations uptake was chosen to be related to the total dissolved cadmium concentration. The development of a litter layer on the soil surface can also be simulated. In the present calculation a layer of 5 kg C m^{-2} was assumed (approximately 8-cm) to build up gradually by decomposition of plant leaves that fall in autumn. After 15 years steady state was reached.

Figure 5 shows the calculated trends of plant uptake and the average cadmium concentration in the litter layer on the soil surface during the first 100 years after ending agricultural use of a podzolic soil in the Netherlands. Figure 6a shows the calculated total dissolved cadmium profile for the same soil after 0, 25, 50, 75 and 100 years for the spring quarter. Figure 6b shows the calculated soil cadmium profile as a function of time. Within 50 years there is a large increase in cadmium in the soil solution, in the litter and in the plants, while the maximum cadmium level is slowly moving down the profile. The increased availability for plants may present a risk to the food chain.

The results illustrate the displacement of basic cations from the soil solid phase by protons. The displacement of cations and cadmium by protons from the soil solid phase, together with the net precipitation, lead to a downward movement of cadmium in the soil solution and to a decrease of soil cadmium. In the present calculation cadmium mobility and availability may be overestimated because no return of basic cations from decomposing leaves, or basic cation accumulation in the rooting zone was assumed. Also the acid neutralizing effect of the process of weathering of minerals, yielding basic cations, was neglected thus far.

Figure 5

**Calculated annual cadmium uptake by plants and
the average annual cadmium level in the litter layer:
preliminary calculations.**

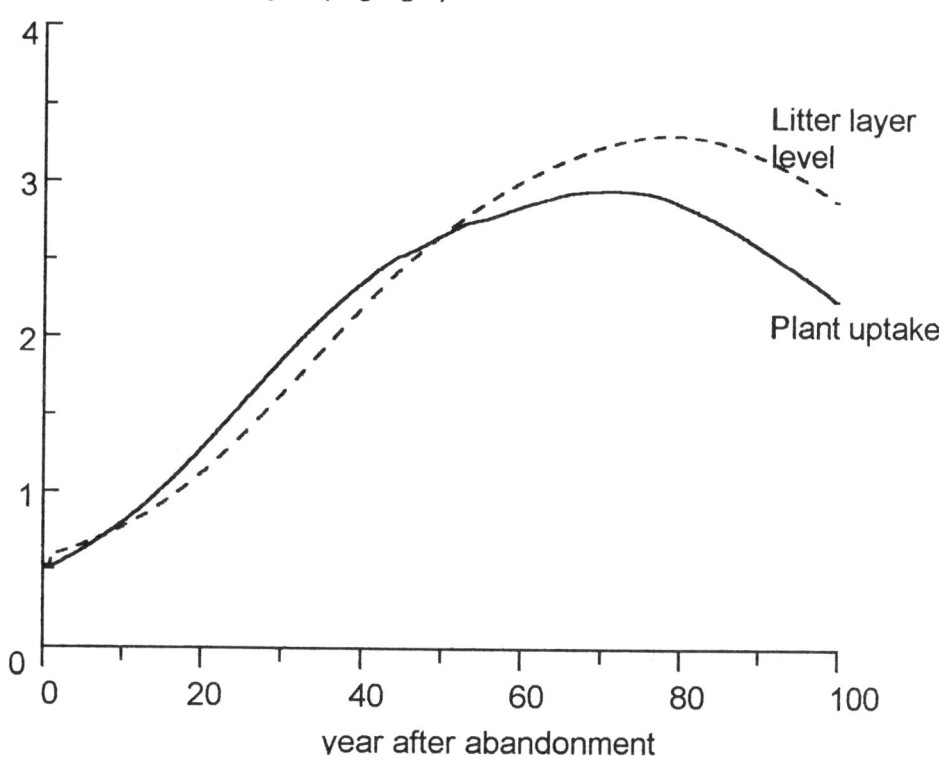

Figure 6

Calculated soil solution cadmium concentration profiles (a); calculated soil cadmium level profiles (b): preliminary calculations.

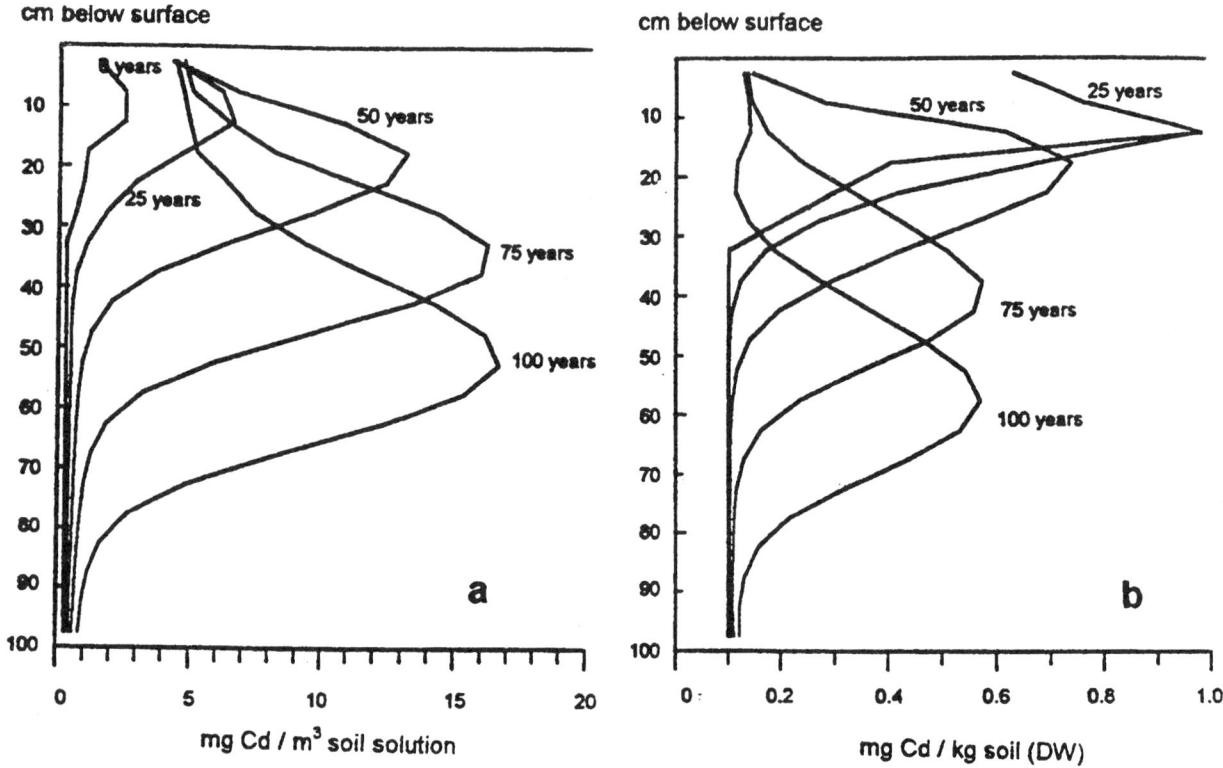

Conclusions

Land use change may have significant implications. Decrease of pH, and the decrease of cadmium complexation at lower pH, will probably make cadmium more harmful to organisms and enhances leaching. However, when a litter layer is formed, the DOC and the organic litter layer both increase the possibility for sorption, while basic cations may be returned to the soil surface when leaves decompose.

The simple soil solution cadmium concentration approach to estimate and predict cadmium uptake by crops yielded promising results, and may be improved when speciation aspects are taken into account. The soil solution cadmium activity approach offers further advantages as it has a sound soil-chemical thermodynamic base. We therefore propose to change the base of the Dutch soil quality criteria (present recommendations) from the soil solid phase to the soil solution.

A reduction of the accumulation rate of cadmium in soil will be achieved by regulations to diminish the cadmium input from various sources. The reduction may roughly be estimated from input and output cadmium balance sheets. For a more precise estimation, when soil acidification continues, also long-term changes in soil chemistry should be taken into account. In the special case that agricultural land is changed into a nature area or woodland, many changes may occur within decades. The changed ecology and chemistry will affect the mobility and availability of cadmium, and must be included in cadmium accumulation calculations. Indicative calculations, taking into account the above factors, predicted major accumulation changes within a period of 50 years. A combined soil chemical/soil ecological model calculation, using sound experimental results, has to be used for realistic balance calculations.

Literature

Bergkvist, B., 1987. Soil solution chemistry and metal budgets of spruce forest ecosystems in S. Sweden. Water Air Soil Pol. 33:130-154.

Bril, J., 1994. Chapter V, Transfer functions between adsorption constants and soil characteristics. In G.J. Reinds, J. Bril, W. de Vries, A. Breeuwsma and J.E. Groenenberg. Critical and present loads of cadmium, copper and lead for European forest soils. SC-DLO/AB-DLO report No. 96. SC-DLO. Wageningen, the Netherlands.

Castilho, P. del, W.J. Chardon and W. Salomons, 1993a. Influence of cattle manure slurry application on the solubility of cadmium, copper, and zinc in a manured, acidic, loamy-sand soil. J. Environ. Qual. 22:689-697.

Castilho, P. del, J.W. Dalenberg, K. Brunt and A.P. Bruins, 1993b. Dissolved organic matter, cadmium, copper and zinc in pig slurry - and soil solution - size exclusion chromatography fractions. Int. J. Environ. Anal. Chem. 50:91-107.

Castilho, P. del and W.J Chardon, 1995. Uptake of soil cadmium by three field crops and its prediction by a pH-dependent Freundlich sorption model. Plant and Soil 171:263-266.

Dalenberg, J.W. and W. van Driel, 1990. Contribution of atmospheric deposition to heavy-metal concentrations in field crops. Neth. J. Agric. Sci. 38:369-379.

Driel, W. van and K.W. Smilde, 1990. Micronutrients and heavy metals in Dutch agriculture. Fert. Res. 25:115-126.

Erp, P. van and P.C. Meeuwissen, 1994. Effect of fertilizer use on the heavy metal supply in Dutch agricultural soils (In Dutch). VROM Report Heidemij advies 674/BA94/A075/07592. VROM. The Hague, the Netherlands.

Langeweg, F. (ed.), 1989. Tomorrow's Concerns. National Environmental Survey 1985-2010 (In Dutch). RIVM. Samson HD Tjeenk Willink, Alphen aan den Rijn, the Netherlands.

Li, Y.M., R.L. Chaney and A.A. Schneiter, 1994. Effect of soil chloride level on cadmium concentration in sunflower kernels. Plant and Soil 167:275-280.

Römkens, P.F.A.M. and W. de Vries, 1995. Acidification and metal mobilization: effects of land use changes on cadmium mobility. In: G.J. Hey and J.W. Erisman (eds.), Acid Rain Research: Do we have enough answers. Elsevier Science, pp. 367-380.

Factors Influencing Cadmium Content in Crops – Results from Swedish Field Investigations

Jan Eriksson, Ingrid Öborn, Gunilla Jansson and Arne Andersson

Department of Soil Sciences
Swedish University of Agricultural Sciences
Uppsala, Sweden

The complete paper presented at the OECD Cadmium Workshop will appear in the Swedish Journal of Agricultural Research, vol. 26 no.3 (1996).

Abstract

In this paper we review results from 20 years of Swedish field studies focused on evaluating the influence of soil cadmium content on cadmium levels in agricultural crops. These investigations have been performed on fields subject to normal agricultural management where phosphorous fertilizers and atmospheric deposition are the main cadmium sources. At present, these sources are roughly of equal size.

Statistical analysis of the sampled material shows that the cadmium content (extractable in boiling 2M HNO_3) of the soil (positive relation) and soil pH (negative relation) are the main soil factors influencing the uptake of cadmium by plants from Swedish soils. In wheat, the soil cadmium level is the most important factor. Uptake is also positively correlated with precipitation during the growing season. Budget calculations show that soil cadmium contents are still increasing. Furthermore, there are areas in Sweden with elevated soil cadmium levels where cadmium contents of 5-10 per cent of the wheat lots produced are near or above limit values (0.1 mg kg^{-1}) for tolerable contents for cereals. Liming has limited value as a measure for decreasing uptake since wheat soils generally have the near neutral pH value considered as optimal for most crops.

We conclude that the level of cadmium in soil is an important and decisive factor determining its uptake and that there are areas where a further increase in soil cadmium levels should be avoided. To further efforts to decrease the input of cadmium to soil we suggest that the declaration of the cadmium contents of all phosphorous fertilizers be made mandatory. It is also necessary to make "cadmium-free" phosphorous fertilizer available on the market. It could be used on soils high in plant-available cadmium and for crops that require large doses of phosphorous fertilizer or that naturally take up large amounts of this metal. A further reduction in emissions of cadmium to the atmosphere is also necessary to get a balance between input and output of cadmium to agricultural soils.

Management Factors Influencing Cadmium Accumulation in Crops

C. A. Grant, L. D. Bailey and W.T. Buckley

Agriculture and Agri-Food Canada, Brandon Research Centre
Brandon, Manitoba, Canada

The accumulation of cadmium by crops is mediated by genetic, environmental, pedological and anthropogenic factors. When considering the effects of management practices on cadmium in the soil-plant system, attention must be paid both to the long-term effects on cadmium accumulation and bioavailability in the soil and the immediate effects on uptake, accumulation and distribution of cadmium in plants. cadmium uptake by plants is influenced by the concentration of cadmium in the soil solution (Mullins et al. 1986) and by plant factors such as production of root exudates. cadmium concentration in the soil solution is a function of the soil pH, cation exchange capacity, soil organic matter, ionic strength of the soil solution, and the presence of competing or complexing ions (Hahne and Kroontje 1973, Doner 1978, Christensen 1984a,b, Eriksson 1990, Naidu et al. 1994). Native soil cadmium concentration and inputs of cadmium into the soil system from natural and anthropogenic sources will also impact on soil solution cadmium concentration. Agricultural management practices which influence soil solution cadmium concentration, either by effects on soil chemical properties or addition of cadmium to the soil systems, may influence cadmium accumulation in crops.

Fertilizer management may influence cadmium concentration in the soil solution, directly and indirectly. Many sources of phosphate, micronutrients and lime contain varying amounts of cadmium which directly increase the cadmium concentration in the soil. The relative importance of this direct addition on long-term accumulation of cadmium in the soil will depend on the concentration of cadmium in the fertilizer or soil amendment and the amount that is applied over time. Long-term field studies have shown that cadmium accumulation in the soil from application of P fertilizers was insignificant (Mortvedt 1987) or measurable (Andersson and Hahlin 1981, Baerug and Singh 1990, He and Singh 1993), although effects on cadmium accumulation in plants were inconsistent. The sensitivity of the soi-plant system to additions of cadmium must also be considered, taking into account the existing soil cadmium content, additions from other sources, and the ability of the soil to adsorb or complex cadmium.

Fertilizer additions can increase cadmium concentration in the soil solution, even if the fertilizer contains no cadmium, through effects on soil pH and ionic strength of the soil solution (Andersson 1976, Garicia-Miragaya and Page 1976, Naidu et al. 1994). Cadmium concentration in crops has been increased by applications of N (Andersson 1976, Eriksson 1990, Oliver et al. 1993, Grant et al. 1995), P (Sparrow et al. 1993, McLaughlin et al. 1995. Grant et al. 1995), and K (Sparrow et al. 1994, McLaughlin et al. 1995, Grant et al. 1995) and either increased (Williams and David 1976) or decreased (Abdel-Sabour et al. 1988, Choudhary et al. 1994) by application of Zn. Fertilization may also influence plant growth parameters, such as root development, which could impact on the ability of the plant to extract cadmium from the soil. Increases in mass flow due to fertilizer applications may increase plant accumulation of cadmium (Lorenz et al. 1994). Long-term acidification from fertilization may be a critical factor in mobilization of soil cadmium for plant uptake (Nicholson et al. 1994).

Fertilizer application techniques may be manipulated to slow the long-term accumulation of cadmium in the soil system. Placement techniques to optimize efficiency of fertilizer use will reduce the total amount of fertilizer applied, minimizing cadmium additions to the system. Use of low-cadmium fertilizer sources, if economically feasible, would reduce total cadmium additions. In terms of short-term input of cadmium into the food chain, selection of fertilizer sources which produce lower plant cadmium concentrations may be effective. Nitrate-N may produce lower plant cadmium concentrations than ammonium-N sources (Eriksson 1990, Florijn et al. 1992). Diammonium phosphate may reduce plant cadmium accumulation as compared to monoammonium phosphate (Levi-Minzi and Petruzzelli 1984) while potassium sulphate may produce lower cadmium accumulation in plants than does potassium chloride (Sparrow et al. 1994). With phosphate fertilizer, plant uptake of cadmium tended to be greater from band placement than from broadcast applications, following a similar pattern to the efficiency of P uptake (Sparrow et al. 1992, 1993). Therefore, use of placements which produce lower cadmium concentration in plants could lead to lower efficiency of P utilization, resulting in requirement for higher fertilization rates to optimize yields, thus increasing the cadmium level in the soil.

Other management factors may also influence cadmium accumulation in plants. cadmium concentration may be higher in crops grown in rotation after other high-accumulating crops such as flax. Return of high cadmium residue to the soil may increase cadmium availability for subsequent crops. Changes in yield potential of crops may impact on cadmium concentration. Healthy, more vigorous plants may have higher rooting and higher transpiration rates, increasing access and absorption of cadmium in the soil. However, increased growth may also lead to dilution of cadmium in the tissue, increasing accumulation but decreasing concentration of cadmium. Tillage management may impact pH and nutrient stratification in the soil profile, which may influence availability of cadmium for plant uptake. The amount of cadmium accumulated by a plant and its distribution among plant parts differs with species (Pettersson 1977) and among cultivars within a species (Pettersson 1977, Chang et al. 1982). Due to the genetic variation in cadmium accumulation, plant breeding programs can develop crop cultivars which tend to accumulate relatively low levels of cadmium in edible parts. Breeding programs are currently underway at Agriculture and Agri-Food Canada in western Canada to produce cultivars of durum and flax with reduced seed cadmium concentrations.

Management strategies to reduce cadmium accumulation in crops should focus on both the immediate and long-term considerations. For "sensitive" crops, with potential for high cadmium transfer into the food chain, practices should include breeding and production of cultivars which tend to accumulate low levels of cadmium, selection of fertilizer sources which encourage lower plant accumulation of cadmium and avoidance of sequencing after high accumulator crops such as flax. To minimize long-term accumulation of cadmium in the soil, strategies could include utilization of minimum levels of fertilizers for optimal crop yields and development of a total loading limit for cadmium on agricultural soils, taking into consideration all anthropogenic sources, the background level of cadmium in the soil and the ability of the soil to adsorb or complex the cadmium. Long-term changes in soil properties, such as pH and organic matter content, must also be considered in view of their potential effects on cadmium availability to crops.

References

Abdel-Sabour, M.F., Mortvedt, J.J. and Kelose, J.J. 1988. Cadmium-zinc interactions in plants and extractable cadmium and zinc fractions in soil. Soil Sci. 145:424-431.

Andersson, A. 1976. On the influence of manure and fertilizers on the distribution and amounts of plant available cadmium in soils. Swedish J. Agric. Res. 6:27-36.

Andersson, A. and Hahlin, M. 1981. Cadmium effects from phosphorus fertilization in field experiments. Swedish J. Agric. Res. 11:3-10.

Baerug, R. and Singh, B.R. 1990. Cadmium levels in soils and crops after long-term use of commercial fertilizers. Nor. J. Agric. Sci. 4:251-260.

Chang, A.C., Page, A.L., Foster, K.W., and Jones, T.E. 1982. A comparison of cadmium and zinc accumulation by four cultivars of barley grown in sludge-amended soils. J. Environ. Qual. 11:409-412.

Choudhary, M., Bailey, L.D., and Grant, C.A. 1994. Effect of zinc on cadmium concentration in the tissue of durum wheat. Can. J. Plant Sci. 74:549-552.

Christensen, T. 1984a. Cadmium soil sorption at two concentrations: I. Effect of time, cadmium load, pH, and calcium. Water, Air, Soil Pollut. 21:105-114.

Christensen, T. 1984b Cadmium soil sorption at two concentrations: II. Reversibility, effect in changes in solute composition, and effect of soil aging. Water, Air, Soil Pollut. 21:115-125.

Doner, H.E. 1978. Chloride as a factor in mobilities of Ni(II), Cu(II), and cadmium(II) in soil. Soil Sci. Soc. Am. J. 42:882-885.

Eriksson, J. 1990 Factors influencing adsorption and plant uptake of cadmium from agricultural soils. Department of Soil Sciences Reports and Dissertations. 4. Swedish University of Agricultural Sciences. Uppsala. 29 pp.

Florijn, P.J., Nelemans, J.A. and van Beusichem, M.L. 1992. The influence of the form of nitrogen nutrition on uptake and distribution of cadmium in lettuce varieties. J. Plant Nut. 15:2405-2416.

Garicia-Miragaya, J. and Page, A.L. 1976. Influence of ionic strength and inorganic complex formation on the sorption of trace amounts of cadmium by montmorillonite. Soil Sci. Soc. Am. J. 40:658-663.

Grant, C.A., Bailey, L.D., Choudhary, M., Brown, K.R. and Racz, G.J. 1995. Fertilizer effects on cadmium concentration in crops. Papers presented at the 38th annual Manitoba Society of Soil Scinece Meeting. Winnipeg. Jan. 3 and 4, 1995. Manitoba Society of Soil Science. pp. 2-10.

Hahne, H.C.H. and Kroontje, W. 1973. Significance of pH and chloride concentration on behavior of heavy metal pollutants: Mercury (11), cadmium (11), Zinc (11) and lead (11). J. Environ. Quality. 2:444-450.

He, Q.B. and Singh, B.R. 1993. Plant availability of cadmium in soils. I. Extractable cadmium in newly and long-term cultivated soils. Acta Agric. Scand. 43:134-141.

Levi-Minzi, R.and Petruzzelli, G. 1984. The influence of phosphate fertilizers on cadmium solubility in soil. Water, Air and Soil Pollut. 23:423-429.

Lorenz, S.E., Hamon, R.E., McGrath, S.P., Holm, P.E. & Christensen, T.H. 1994. Application of fertilizer cations affect cadmium and zinc concentrations in soil solutions and uptake by plants. Eur. J. Soil Sci. 45:159-165

McLaughlin, M.J., Maier, N.A., Freeman, K., Tiller, K.G., Williams, C.M.J., and Smart, M.K. 1995. Effect of potassic and phosphatic fertilizer type, fertilizer cadmium concentration and zinc rate on cadmium uptake by potatoes. Fertilizer Res. 40:63-70.

Mortvedt, J.J. 1987. Cadmium levels in soils and plants from some long term fertility experiments in the United States of America. J. Environ. Qual. 16:137-142.

Mullins, G.L., Sommers, L.E. and Barber, S.A. 1986. Modelling the plant uptake of cadmium and zinc from soils treated with sewage sludge. Soil Sci. Soc. Am. J. 50:1245-1250.

Naidu, R., Bolan, N. S., Kookana, R.S. and Tiller, K.G. 1994. Ionic-strength and pH effects on the sorption of cadmium and the surface charge of soils. Eur. J. Soil Sci. 45:419-429.

Nicholson, F.A., Jones, K.C. and Johnston, A.E. 1994. Effect of phosphate fertilizers and atmospheric deposition on long-term changes in the cadmium content of soils and crops. Environ. Sci. Technol. 28:2170-2175.

Oliver, D.P., Schultz, J.E., Tiller, K.G. and Merry, R.H. 1993. The effect of crop rotations and tillage practices on cadmium concentration in wheat grain. Aust. J. Agric. Res. 44:1221-34

Pettersson, O. 1977. Differences in cadmium uptake between plant species and cultivars. Swedish. J. Agric. Res. 7:21-24.

Sparrow, L.A., Chapman, K.S.R., Parsley, D., Hardman, P.R. and Cullen, B. 1992. Response of potatoes (Solanum tuberosum c.v. Russet Burbank) to band-placed and broadcast high cadmium phosphorus fertilizer on heavily cropped krasnozems in north-western Tasmania. Aust. J. Exper. Agric. 32:113-119.

Sparrow, L.A, Salardini, A.A. and Bishop, A.C. 1993. Field studies of cadmium in potatoes (Solanum tuberosum L.). II. Response of cvv. Russet Burbank and Kennebec to two double superphosphates of different cadmium concentration. Aust. J. Agric. Res. 44:855-61.

Sparrow, L.A, Salardini, A.A. and Bishop, A.C. 1994. Field studies of cadmium in potatoes (Solanum tuberosum L.). III. Response of cv. Russet Burbank to sources of banded potassium. Aust. J. Agric. Res. 45:243-9.

Williams, C.H. and David, D.J. 1976. The accumulation in soil of cadmium residues from phosphate fertilizers and their effect on the cadmium content of plants. Soil Sci. 121:86-93.

Uptake of Cadmium by Crop Plants

L.D. Bailey, C.A. Grant and R.J. Hill

Joint Position Paper (Agriculture and Agri-Food Canada and Health Canada)

Cadmium Bioavailability and Crop Plant Uptake and Accumulation:

The natural abundance of cadmium in the agricultural environment is relatively small, ranging from 0.1 to 0.4 mg/kg soil but can be as high as 4.5 mg/kg in volcanic soils. But anthropogenic sources can significantly increased the total cadmium content of some agricultural soils, thus increasing the potential for cadmium to enter the food chain as a contaminant and possible health threat. Inputs of cadmium to soils can be classified either as **primary sources** if they are a result of agricultural practices such as the direct application of phosphate fertilizers, manures and sewage sludge or **secondary sources** if the cadmium is deposited from the atmosphere following emission from the mining and metallurgical industries, forest fires and volcanic activities.

In contaminated soils cadmium values can range from 6 to 97 mg/kg, with concentrations recorded as high as 600 mg/kg. The EEC standard for agricultural soils is 1 to 3 mg/kg, but in some contaminated EC soils mean values of 1 to 10 mg/kg have been reported. Knowledge of the total content of cadmium in soils provides only limited information when considering its toxic effect, since total cadmium content of soils does not correlate well with biological availability and gives no information on the chemical reactivity of the different forms of the metal found in soils (Spevackova and Kucera, 1989). Some measure of bioavailability and mobility is required if reliable evaluations of pollution hazards are to be made.

The soil solution plays a critical role in determining the availability of metal ions to plants. Both the solubility and bioavailability of heavy metal ions are affected by a number of factors (Lorenz et al., 1994). For metals in general, besides the addition of fertilizers, manures and other amendments, soil pH is the most important factor (Sanders et al., 1986; Chang et al., 1987; Anderson and Christensen, 1988; Basta and Tabatabai, 1992a). Other important factors include clay content, oxides of Mn and Fe (King, 1988), ionic interactions (Basta and Tabatabai, 1992b), redox potential (Brown et al., 1989) organic matter (McGrath et al., 1988) and the plant species itself (Treeby et al. 1989; Zang et al., 1991). Needless to say, understanding of the bioavailability of cadmium in soils is still at the "controversial" stage of scientific progress. In both Europe and in North America research in this field continues to be a high priority.

Effect of plant genetics, macro and micro nutrients

The amount of cadmium accumulated in plant and subsequently translocated to the seed is influenced by several factors including plant genetics (William and Davis, 1973, Leisle and Penner, personal communication, Kenaschuk and Dribnenki, personal communication), cadmium concentration of the soil solution (Turner, 1973), soil temperature (Sheaffer et al., 1976), and management practices (Eriksson, 1990; Singh, 1990, Grant et al., 1995). There are wide variations among plant species and cultivars in their ability to take up, accumulate, and translocate cadmium to seeds. For example some crop species such as durum wheat, flax seed and sunflower seed have levels of cadmium higher than 0.1 ppm, the Codex proposed limit, while spring wheats, barleys, oats, canolas and corns, have levels of cadmium < 0.1 ppm. Leisle (AAFC, Winnipeg RC, personal communication) reported significant variation in the cadmium content of durum wheat cultivars, grown under the same environment. In more recent research, Leisle and Penner (AAFC, Winnipeg RC, personal communication) reported the isolation of a gene in durum wheat that segregates for low cadmium. Using the "marker gene" they screened and isolated durum lines with cadmium content 30 to 60 per cent below the high cadmium lines. Currently all Canadian durum lines in registration tests have cadmium levels significantly lower than the proposed Codex level of 0.1 ppm. In studies with oilseed flax, Kenaschuk and Dribnenki (AAFC, Morden RC and United Grain Growers, respectively, personal communication) reported great diversity in their population of flax seed. They anticipate a "break through" with flax similar to that obtained by Leisle and Penner with durum. In a USA study to identify sunflower genotypes with low cadmium uptake and transportation to kernels, Chaney et al. (1993), reported that among the genotypes tested, there was sufficient evidence to believe that the potential for low cadmium genetic lines exist. Genetic manipulation to develop crop cultivars that segregate against cadmium and other heavy metal uptake and accumulation is a promising tool in solving the problem of unwanted heavy metals in crops.

Application of macro- and micro- nutrients has been shown to impact on cadmium concentration in a variety of crops (Eriksson, 1990; Choudhary et al., 1994, 1995: Grant et al., 1995). Phosphate fertilizers have been shown to increase cadmium uptake of wheat (Andersson and Siman, 1991) or to have no significant effect on crop uptake (Choudhary et al., 1994, 1995; Grant et al., 1995; Mortvedt, 1987). The effect of zinc on uptake and accumulation of cadmium by plants has been extensively studied. The application of $ZnSO_4$ has been shown to be antagonistic to cadmium uptake (Abdel-Sabour et al., 1988, Choudhary et al., 1994, 1995), have no effect (White and Chaney, 1980), or be synergistic (Williams and David 1976). Choudhary et al. (1995) found that when zinc was applied to the soil, the level od cadmium in the grain, stem, and root of durum wheats were reduced, whereas foliar applied zinc had no similar effect. Soil applied zinc was particular effective in reducing cadmium levels in plant parts when it was applied with nitrogen and phosphorousfertilizers, materials that were found to increase the cadmium concentration of the crop. These results are in agreement with those of Jarvis et al. (1976) who reported that zinc may depress the uptake of cadmium by competing for exchange sites at the root surface. Christensen (1987) found that while of Zn, Ca, Co, Cr, Cu, Ni and Pb can inhibit cadmium uptake, it was zinc that had the greatest inhibitory effect. Grant et al. (1995) also found that the addition of zinc to soils depressed cadmium uptake of durum and flax, primarily when they were low or deficient in zinc for optimum crop growth. The depressing effect of cadmium uptake observed in pot experiments was due to the tendency towards zinc deficiency which is often observed in pot experiments due to restricted root growth in a limited volume of soil. Chaney et al. (1994) arrived at similar conclusions in studies in the USA, and presented data to show a four-fold increase in cadmium concentration in wheat grown on zinc deficient soils.

In studies on the Canadian prairies, durum grain yield and cadmium content varied with soil type, fertilizer placement (nitrogen and phosphorous), and with year, reflecting the variation in growing conditions. The source of phosphorous -fertilizer had no effect on cadmium uptake, but, addition of seed placed phosphorous-fertilizer significantly increased the cadmium content of the grain compared to banding the phosphorousaway from the seed. On Black Chernozemic soils, Grant et al. (1995) found that application of monoammonium phosphate to durum wheat increased the cadmium content in the grain. However, the level of accumulation did not differ significantly when the phosphorous source was reagent grade or commercial fertilizer grade. Consequently, the observed increase in cadmium in the grain was not due only to the presence of cadmium in the P-fertilizer, but may have been the results of other soil and plant factors or due to the nitrogen component of the fertilizer. The authors also reported that the application of $ZnSO_4$ with phosphorous-fertilizer, reagent or commercial grade, significantly reduced the cadmium content of durum wheat cultivars, while the addition of urea-nitrogen fertilizers, which contain no cadmium, increased cadmium uptake. It is apparent therefore, that factors that stimulate plant growth, such as fertilization, optimum water and temperature may stimulate cadmium uptake.

Research requirements

The development of effective and meaningful agricultural strategies for reducing the cadmium content of food crops will depend on as complete an understanding as possible of the numerous factors that influence cadmium uptake and accumulation. Accordingly, the areas of research that should be actively pursued should include: (i) the entry, accumulation, soil stratification and bioavailability of soil-cadmium; (ii) the impact of tillage systems, crop sequencing and other agronomic management practices on the uptake and accumulation by crops; (iii) the effect of sewage sludge and other organic soil amendments (manures, composted wastes, etc.) on soil and crop cadmium levels; (iv) an elucidation of the biochemical mechanisms of cadmium uptake, translocation and accumulation within plants and the linkage of these processes to differences in the genetic makeup of plants; (v) genetic modification of current crop cultivars to reduce cadmium uptake and accumulation; (vi) the relative importance of atmospheric deposition of cadmium levels in soils and crops. Furthermore, much research needs to be carried out before we will be able to make definitive statements about the relative bioavailability and toxicity of various species of cadmium, in particular cadmium bound to plant and animal metallothioneins.

Risk management options

Amongst the risk management options that should be considered are the recommendation of a total loading limit for cadmium on agricultural soils. This limit should include all anthropogenic sources (e.g. fertilizers, sewage sludge, manures, atmospheric deposition,etc.), but also take in to consideration the total background level of the cadmium in the soil. For such action to be meaningful, it would be necessary to ascertain the natural background level of cadmium in the soil, and desirable to establish standard methods for the analysis of cadmium in soils and other samples.

Removal of cadmium from natural occurring sources is expensive and may not contribute significantly to solving the problem of cadmium transfer in soil-plant-food system. I such technology is adapted it is important to ensure that: (i) there is no increase in environmental pollution associated with the process, and (ii) once removed, the cadmium containing wastes are disposed of safely. A more promising approach to the reduction of human exposure would be to develop technologies for the reduction of cadmium uptake by plants from natural sources as soil and fertilizers.

References

Abdel-Sabour, M. F., Mortvedt, J. J., and Kelose, J. J. 1988. Cadmium-zinc interactions in plants and extractable cadmium and zinc fractions in soil. Soil Sci. 145(6): 424-431.

Anderson, P.R. and Christensen, T.H. 1988. Distribution coefficients of cadmium, Co, Ni, and zinc in soils. J. Soil Sci. 39: 15-22.

Andersson, A., and Bingefor, S. 1985. Trends and annual variations in cadmium concentrations in grain of winter wheat. Acta Agric. Scanda. 35: 339-344.

Andersson, A. and Siman, G. 1991. Levels of cadmium and some other trace elements in soils and crops as influenced by lime and fertilizer level. Acta. Agric. Scand. 41: 3-11.

Basta, N.T. and Tabatabai, M.A. 1992a. Effect of cropping systems on adsorption of metals by soils: II. Effect of pH. Soil Sci. 153: 195-204.

Basta, N.T. and Tabatabai, M.A. 1992b. Effect of cropping systems on adsorption of metals by soils. III. Competitive adsorption. Soil Sci. 153: 331-337.

Brown, P.H., Dunemann, L., Schulz, R. and Marschner, H. 1989. Influence of redox potential and plant species on the uptake of nickel and cadmium from soils. Zeitschrift für Pflanzenernährung und Bodenkunde. 152: 855-91.

Chang, A.C., Page, A.L. and Warneke, J.E. 1987. Long term sludge application on cadmium and zinc accumulation in swiss shard and radish. J. Environ. Qual. 16: 217-221.

Chaney, R.L., Li, Y-M., Schneiter, A.A., Green, C.E., Miller, J.F. and Hopkins, D.G. 1993. Progress in developing technologies to produce low cadmium concentration sunflower kernels. In Proc. 15th Sunflower Res. Workshop, Fargo, ND.

Choudhary, M., Bailey, L.D. and Grant, C.A. 1994. Effect of zinc on cadmium concentration in the tissue of durum wheat. Can. J. Plant Sci. 74: 549-552.

Choudhary, M., Bailey, L.D., Grant, C.A. and D. Leisle. 1995. Effect of zinc on the concentration of cadmium and zinc in plant tissue of two durum wheat lines. Can. J. Plant Sci. 75:445-448.

Christensen, T.H. 1987. Cadmium soil sorption at low concentrations. V. Evidence of competition by other heavy metals. Water, Air, and Soil Pollut. 34: 305-314.

Eriksson, J.E. 1990. Factors influencing adsorption and plant uptake of cadmium from agricultural soils. Reports and Dissertations-Swedish University of Agricultural Sciences, Dept. of Soil Science. Vol. 4. 29 pp.

Grant, C.A., Bailey, L.D., Choudhary, M., Brown, K.R. and Racz, G.J. 1995. Fertilizer effects on cadmium concentration in crops. In Proc. Manitoba Soc. of Soil Sci., Annual Meeting, U of Man., Winnipeg.

Jarvis, S.C., Jones, L.P.H. and Hopper, M.J. 1976. Cadmium uptake from solution by plants and its transport from roots to shoots. Plant Soil. 44:179-191.

King, L.D. 1988. Effect of selected soil properties on cadmium content of tobacco. J. Environ. Qual. 17: 215-255.

Lorenz, S.E., Hamon, R.E., McGrath, S.P. Holm, P.E. and Christensen, T.H. 1994. Applications of fertilizer cations affect cadmium and zinc concentrations in soil solutions and uptake by plants. European J. Soil Sci. 45: 159-165.

McGrath, S.P., Sanders, J.R. and Shalaby, M.H. 1988. The effects of soil organic matter levels on soil solution concentration and extractabilities of mangenese, zinc and copper. Geoderma. 42: 177-188.

Mortvedt, J. J. 1987. Cadmium levels in soils and plants from some long term soil fertility experiments in the United States of america. J. Environ. Qual. 16:137-142.

Sanders, J.R., McGrath, S.P. and Adams, T. McM. 1986. Zinc, copper and nikel concentrations in ryegrass grown on sewage sludge-contaminatted soils of different pH. J. Sci. Food and Agric. 37: 961-968.

Sheaffer, C.C., Decker, A.M. and Chaney, R.L. 1976. Effect of soil temperature and sludge application on heavy metal content of corn. Agron. Abstr. Am. Soc. of Agron., Madison, Wisconsin, p. 32.

Singh, B.R. 1990. Cadmium and floride uptake by oats and rape from phosphate fertilizers in two different soils. Norw. J. Agric. Sci. 4: 239-249.

Spevackova, V., and Kucera, J. 1989. Trace elemrnt speciation in contaminated soils studied by atomic absorption spectrometry and neutron activation analysis. Int. J. Environ. Anal. Chem. 35: 241-251.

Treeby, M., Marschner, H. and Römheld, V. 1989. Mobilization of iron and other micronutrient cations from calcareous soil by plant-borne, microbial, and synthetic metal chelators. Plant and Soil. 114: 217-226.

Turner, M. A. 1973. Effects of cadmium treatment on cadmium and zinc uptake by selected vegetable species. J. Environ. Qual. 2(1): 118-119.

Williams, C. H. and David, D. J. 1973. The effect of superphosphate on the cadmium content of soils and plants. Aust. J. Soil Res. 11(1):43-56.

Williams, C. H. and David, D. J. 1976. The accumulation in soil of cadmium residues from phosphate fertilizers and their effect on the cadmium content of plants. Soil Sci. 121(2): 86-93.

White, M.C. and Chaney, R.L. 1980. Zn, cadmium and manganese uptake by soybean from two zinc and cadmium-amended costal plain soils. Soil Sci. Soc. Am. J. 44: 308-313.

Zang, F., Römheld, V. and Marschner, H. 1991. Release of zinc mobilizing root exudates in different plant sppecies as affected by zinc nutritional status. J. Plant Nutrit. 14: 675-686.

The Relation Between the Cadmium Content in Soil and in Food Plants

Kimmo Louekari

Finnish Institute of Occupational Health

1. The characteristics of the soil and the cadmium input explain the different observations in different OECD Member countries

The important concern regarding cadmium is its slow accumulation to agricultural soil and consequent increase of the cadmium content in food plants. Among the numerous studies dealing with the soil-plant relationship of cadmium, it is difficult to find firm and consistent patterns or correlation. But what are the reasons behind the somewhat contradictory observations from different countries, reviewed in the OECD Cadmium Monograph and in other literature?

The major factors which determine the plant uptake of cadmium are:

- input of cadmium to soil (see Table 1 and 2);

- soil pH;

- cation exchange capacity (surface area of particles) in soil (Seuri 1990);

- content of organic material in soil;

- plant species and cultivar, (plant part).

The cadmium concentration in fertilizers, application rate of fertilizers, sewage sludge application, and atmospheric deposition vary among countries and agricultural areas. Soils pH, cation exchange capacity and content of organic material in soil affect the solubility and bioavailability of cadmium (Alloway et al. 1988). Obviously, the agricultural soils in the OECD Member countries differ in respect to these factors, and the cadmium concentration in soil is only one determinant to be considered.

As an example, when analysing some trace elements in soil and rice in Catalonia, Schuhmacher et al. (1994) found that the cadmium content was higher in peat and clay soils than in sandy soils, whereas no significant difference was observed in cadmium concentrations in rice grown in various soils. This may indicate that a high cadmium content in clay soil is tolerable, since in that type of soil, cadmium is less available to plant uptake. Referring to Suzuki et al. (1980) the same authors discuss, that pH of the soil, content of humus, formation of insoluble compounds and oxidation-reduction compounds are important determinants of the plant cadmium content and conclude that only when these conditions are constant, cadmium-rich soil produces cadmium-rich rice.

2. Estimates on long-term increase of soil and plant cadmium content

Calculations and trend studies in Sweden, Denmark and the United Kingdom (Andersson et al. 1985, Tjell et al. 1981, Jones et al. 1989) have shown that a slow increase of cadmium content in soil may result in an increased cadmium content in cultivated plants (Table 1). In the two long-term studies available, one from Uppsala, Sweden, and another from Rothamstedt, UK, increase of plant cadmium has been observed. In the Rothamstedt study, the importance of the content of organic material in soil was revealed, when the effect was seen in areas treated with NPK fertilizers, whereas that was not the case for areas treated with sewage sludge.

In some other studies, probably due to the low cadmium input rate, relatively high soil pH, or other treatments, no effect on plant cadmium content has been observed. In the study reported by Mäkelä-Kurtto the total cadmium input to Finnish soils was relatively small, since the imported cadmium-rich raw phosphate was used for less than ten years and was gradually substituted by very low apatite in early eighties. However, the average soil cadmium level in Finland was increased after imported phosphate fertilizers had been used for some years (Mäkelä-Kurtto 1989). Later, the lower cadmium content in fertilizers and decreased atmospheric emissions resulted in a balance of cadmium in Finnish agricultural soils (Mäkelä-Kurtto 1994).

The calculated annual accumulation of cadmium in soils in Denmark is 4g/ha, which would increase the soil concentration in average by about 0.6 per cent annually (Tjell et al. 1981). It was estimated, that this would result in almost doubling of the dietary intake of cadmium in Denmark from 30 to 50 µg/day. Inflow from fertilizers was 3.0 and from atmospheric deposition 2.0 g/ha/year. Outflow with drainage water is 0.2-0.5 g/ha.

The crop uptake of cadmium in Denmark has been assessed to be in the order of 1.1 g/ha (Hovmand 1980). However, 0.95 is assessed to be recycled back with manure (Hovmand). According to Christensen the absorption of cadmium in topsoil is almost fully reversible and apparently no appreciable fixation takes place during ageing. Using isotope dilution techniques it has been revealed that about 30 per cent of cadmium in grass crop originate directly from atmosphere (Tjell 1981). The atmospheric deposition has been reduced also in Denmark since 1980 (OECD 1994), which would result in slower accumulation of cadmium in soils.

It is noteworthy that in some long-term studies and surveillance in USA and Australia, no noticeable increase in plant cadmium was observed, although the soil cadmium content elevated (Mortwedt et al. 1987, Gartrell 1990). The follow-up reported by Mortwedt shows that when the cadmium concentration of fertilizers and annual cadmium input are at low level, no effect on plant cadmium is seen, though the soil cadmium content increased (Table 1). In Europe, however, the atmospheric deposition of cadmium is probably higher and the amount of phosphate fertilizers per hectare greater than in North America. In USA, as noted by Mortvedt (1987), also the cadmium content of fertilizers is lower (often at the level of 5 mg/kg) than in Europe. In Scandinavia, the prevailing soil types are relatively acidic and have low cation exchange capacity, which enhances the plant cadmium uptake.

There are very few long-term studies with a sampling period of 50 - 100 years. More trend data would be valuable for assessments the risk of cadmium in agricultural soil. It is likely that sample archives would contain e.g. soil and crop samples representing areas which have faced various inflows of cadmium via fertilizers, sewage sludge, manure and atmospheric deposition. There are serious limitations when interpreting the current data, especially concerning the effect of long-term cadmium input to edible plants in various soil conditions. To improve possibilities of meaningful assessment, trend studies should include all relevant parameters listed in chapter 1. At present, it remains questionable whether results, which describe trends in some area or country are relevant when concerning other agricultural conditions.

The annual amount of fertiliser application as such tells very little about the input of cadmium to the soil, since cadmium content of fertilizers varies. Fertilizers of West African origin contain 160-255 mg of cadmium/kg, whereas fertilizers derived from south-eastern USA contain 35 mg/kg (Hutton 1982), and cadmium concentration in fertilizers made of mineral apatite in Finland is 0.5-1 mg/kg (Louekari et al. 1994).

3. Contaminated areas and field experiments

Numerous contaminated areas in the OECD Member countries have been studied, and often a considerable increase in plant cadmium content has been reported (Table 2). Such high levels of contamination summarized in Table 2, probably and hopefully never apply with the vast majority of agricultural lands. Data that has accumulated from the contaminated areas and from experiments might, however, help to predict the soil-plant relationship in general. Again, any prediction would be complicated, since the important characteristics of the soil are seldom reported.

When 800 g of cadmium/ha was added to soil, the vegetable cadmium content was increased to 2-3 fold. The amount of cadmium was added in form of municipal sewage sludge and crops were grown in calcareous soil. Assuming the original cadmium content of 600 g of cadmium/ha of soils, the addition increased the soil cadmium content respectively, that is 2-3 fold (Page et al. 1982). What makes this experiment incomparable to actual field conditions is that high amount of cadmium was added once, whereas normally cultivated soil receives small amount of cadmium annually with fertilizers and from atmosphere. The effect of pH was also elucidated by Page et al. (1982). It was shown that cadmium concentration of pea, wheat and barley increased about 3-fold, when pH decreased from 7.3 to 4.6. In this experiment 1200 g of cadmium/ha was added in municipal sewage sludge.

Muller et al. (1994) have studied effect of releases of cadmium from pigment plants on surrounding gardens at Bad Liebenstein. Due to emissions from 1960 to 1988, the soil cadmium content increased from 0.5-1.0 mg/kg to 5.4-20.1 mg/kg in air-dry soil and the cadmium content in lettuce and parsley increased 6- and 9-fold (from 352 µg/kg to 2140 µg/kg and from 136 to 1194 µg/kg in dry matter) (Muller et al. 1984).

In vegetables (cabbages and carrots) grown in polluted soils from Shipham and other polluted areas, cadmium concentration was 10-100 times higher as normally in the UK. In these pot experiments, the soil cadmium content was elevated from 1 µg/g up to 365 µg/g in dry matter, and thus a correlation, although not proportional is seen also in this study (Alloway 1988).

4. The relationship between the cadmium content in soil and in food plants has been proportional in some European studies

In the light of some studies conducted in Europe, the correlation of the soluble cadmium in soils and the cadmium concentration in plants seems proportional. When the cadmium content of soil doubled, this happens also in plants (Naturvårdsverket 1987, Page et al. 1982). When soil cadmium concentration is 1-5 mg/kg (about ten times higher than in uncontaminated soil) the plant cadmium content is 0.2-1.0 mg/kg (wheat, lettuce, celery, carrots), which is about ten-fold compared to the average concentrations (Juste 1992, Houtmeyers 1985, Muller 1994). In some other studies, such relationship is not seen, possibly because of soil conditions and/or very slow annual input of cadmium.

When unfavourable soil condition prevail, i.e. pH and cation exchange capacity are low, it is possible that in 50-100 years the content of wheat will increase two-fold in those agricultural areas, which receive 10-20 g of cadmium/hectare with fertilizers and with atmospheric deposition (Seuri 1990). Addition of cadmium to soils at rates from 0.5 to 2.0 g/hectare does not appear to result in increased cadmium-levels in plants (Mortwedt 1987, Andersson 1977).

It has also been proposed that the content of **soluble** cadmium in soil is more relevant than the total soil cadmium concentration in terms of uptake by plants (Naturvårdsverket 1987, Seuri 1990). Unfortunately, the total cadmium content and not the soluble part of cadmium has been reported in several important studies. Due to the cation binding capacity of the soil, all the added cadmium is not soluble. In different countries and agricultural areas binding capacity in soil (e.g. clay content) varies remarkably (Kabata-Pendias and Pendias 1984). If soil pH decreases by one unit, cadmium content of most cultivated plants increases two- or three-fold (Naturvårdsverket 1987).

5. Conclusions

1. In some countries and agricultural areas, normal fertiliser application rate together with atmospheric deposition seem to increase the cadmium concentration of plants. In some other studies such effect has not been observed, which is probably due to relatively slow increase of soil cadmium, high content of organic material in soil (addition of sewage sludge) or low cadmium concentration in fertilizers.

2. Soil type, pH and cation exchange capacity are not always given, when results on soil-plant relationship are reported. The cadmium concentration in fertilizers, cadmium input to soils as g/ha and original cadmium content of soil are also valuable information, which is missing in some reports. Therefore, results reported from various agricultural areas are difficult to compare and explain.

3. Before the purification of phosphate fertilizers has been technically and economically solved, the concern caused by cadmium accumulating to agricultural soils is, in principle, a global issue. However, the national and local risk of elevated plant cadmium concentrations due to fertiliser application should be evaluated taking into account the factors listed in Section 1.

4. Uptake of cadmium by plant could be diminished in many ways: decrease of the amount of phosphorus fertiliser applied, decrease of cadmium content in fertilizers, liming of soils and increasing of organic material of soils, change of plant species or cultivar. Liming and change of plant species are temporary solutions. The preventive and permanently efficient measure, if fertiliser application can not be reduced, is to diminish the cadmium concentration in the fertilizers used.

5. Whereas the need to control cadmium input to agricultural soils in most European countries is obvious, immediate action in some other countries may not be necessary. Also in these countries, however, the soil cadmium content continues to increase in such agricultural areas where the annual input is more than 5.0 g/hectare. Therefore, and because international trade aspects, concerted action in the OECD forum to control the cadmium balance in soils is justified.

Acknowledgements

Discussions and valuable comments received from Dr. Ritva Mäkelä-Kurtto (Agricultural Research Centre, Finland) during the drafting of this article are gratefully acknowledged.

Table 1

Trend of cadmium concentration in soil and in plant under normal agricultural conditions

Country/Area (Ref)	Input of cadmium to soil Source of cadmium	Increase of cadmium concentration in soil	Effect on plant cadmium	Remarks
Sweden (Andersson et al. 1985)	About 7 g/ha in 80ties 4 g from fertilizers and 3 g from atmospheric deposition pH: 5.0-7.0 (Naturvårdsverket 1987)	Soil cadmium increased by about 430 g/ha, i.e. 70 per cent (estimated by the author)	From 1918 to 1980, concentration of winter wheat (Ultuna variety) increased 2-fold, from 25 to 56 ug/kg (f.w.)	It is proposed that acid precipitation and reduced use of manure may have enhanced the plant cadmium uptake
United Kingdom Rothamstedt (uncontam.) (Jones et al. 1987)	1.9-5.4 g/ha annually (estimated) Superphosphate and rock phosphate Cadmium content were 3.3-40 mg/kg and 3.6-92 (mean 36) mg/kg, resp. Also atmospheric input, manure	Soils cadmium increased by 27-55 per cent since 1850	Not analysed	Most superphosphates used in UK contain much higher cadmium concentration than those used in Rothamstedt (Marocco, Senegal, Tunisia)
United Kingdom Rothamstedt (Jones et al. 1989)	1.9-7.2 g/ha annually (estimate) Mainly form atm. deposition Also from guano P and manure after 1905	Soil cadmium increased by 27-159 per cent since 1850	Increase of grain cadmium was obs. when NPK fertilisers were used but not in areas treated with manure. Cadmium in wheat and herbage increased by about 50 per cent (1860-1985)	Soil pH (5.2-6.5) had considerable effect on the cadmium in herbage
USA Whole country Nine experimental plots (Mortvedt 1987)	0.3-1.2 g/ha annually with P fertilizers Fertilizers cadmium content 3-10 mg/kg	Annual increase of total cadmium was about 0.4 per cent in over 40 or 50 years	Cadmium content of soybean, wheat and timothy did not increase	
Iceland Field trials (Torsteinsson et al. 1984)	39 kg P/ha was applied	Not analysed	Cadmium content of grass increased from 0.04 to 0.073 in one and from 0.025 to 0.031 mg/kg in another area.	Concentration of cadmium in livers of lambs and kidneys of cows were low as compared to permitted levels
Finland Whole country (Mäkelä-Kurtto 1989)	5-10 g/ha annually in 1974-1983 (input of P: 30 kg/ha a) 15-24 mg/kg in rock phosphate	Soil cadmium increased 30 per cent from 0.06 to 0.08 mg/kg (soluble cadmium)[1]	No effect on cadmium in timothy grass	

1) AAAc-EDTA extraction, measure of soluble, bioavailable cadmium

Table 2
Cadmium concentration in oil and in some contaminated agricultural areas and in experimental plots

Country/Area (Ref)	Input of cadmium to soil Source of cadmium	Increase of cadmium concentration in soil	Efect on plant cadmium	Remarks
Germany Bad Liebenstein (Muller et al. 1994)	Pigment factory	Soil cadmium increased 10-20 fold	Cadmium in lettuce and parsley increased 6-9 fold	Cadmium in hair and blood of inhabitants were within the normal range
USA Experimental plots Page 1981	800 g/ha in sewage sludge One treatment/addition	Assuming that soil cadmium content was 600 g/ha, the addition was about 130 per cent	Cadmium content of carrot and radish (foliage and roots, Swiss chard and lettuce (foliage) increased 46-340 per cent (nt)	
USA New York State (Mortvedt 1984)	200 g/ha in 9 years with super-phosphate (TSP)	Soil cadmium increased by about 25 per cent (estimated)	Cadmium content of snap bean seed, beet, grain, cabbage and sweet corn did not increase	The annual TSP application was 16 fold as compared to "optimum" P rate
Finland Harjavalta (Sippola et al. 1986)	Copper refinery Atm. release has been 2-4 tonnes annually	Soil cadmium increased from 0.06 to 0.5 mg/kg (soluble cadmium)[1]	Cadmium in lettuce and carrot increased about 2-fold (from 50 to 90, and from 30 to 62 ug/kg fresh weight, resp.)	
Norway Sulitjelma (Lobersli et al. 1988)	Cadmium input/releases unknown Copper smelter	Soil cadmium increased 3-8 fold From 0.3 to 2.4 mg/kg as total soil cadmium	Positive soil-plant correlation was found for Betula pubescens and Vaccinium myrtillus	Surface deposition is as important contributor as soil cadmium
Iran pot experiment (Shariatpanahi et al. 1986)	2800 g/ha during the study		Cadmium content in vegetables and herd increased 50-300 fold	The cadmium input is comparable to that applied with wastewater used for irrigation. Due to phytotox. the yield of plants decreased by about 50 per cent.

1) AAAc-EDTA extraction, measure of soluble, bioavailable cadmium

References

Alloway, B.J., Thornton, I., Smart, G.A., Sherlock, J.C. and Quinn, M.J. (1988) Metal availability. Sci. Total Environ., 75: 41-49.

Andersson, A. and Bingefors, S. (1985) Trends and annual variations in cadmium concentrations in grain of winter wheat. Acta Agric. Scand. 35: 339-344.

Naturvårdsverket (1987) Kadmium i miljön. Bedömningsgrunder., Rapp. 3317: 1-76.

Davis, R.D. and Coker, E.G. (1980) Cadmium in agriculture, with special reference to the utilisation of sewage sludge on land, Medmenham, United Kingdom, Water Research Centre (Technical Report TR/139).

Jones, K.C., Symon, C.J. and Johnston, A.E. (1987) Long-term changes in soil and cereal grain cadmium: studies at Rothamsted Experimental Station. In: Hemphill, D.D., ed. Trace substances in environmental health - XXI. Proceedings of the University of Missouri's 21st Annual Conference on Trace Substances in Environmental Health, St. Louis, Missouri, 25-28 May, 1987, Columbia, Missouri, University of Missouri Press, pp. 450-460.

Juste, C. and Mench, M. (1992) Long-term application of sewage sludge and its effects on metal uptake by crops. In: Adriano, D.C., ed. Biogeochemistry of trace metals, Boca Raton, Florida, Lewis Publishers.

Louekari, K., Saarikoski, H. and Joki-Kokko, E. (1991) Kadium ympäristössä. Helsinki (abstract in English).

Mortwedt, J.J. (1987) cadmium levels in soils and plants from some long-term soil fertility experiments in the United States of America. J. Environ. Qual 16: 137-142.

Muller, M. and Anke, M. (1994) Distribution of cadmium in the food chain (soil-plant-human) of a cadmium exposed area and the health risks of the general population. Sci. Total. Environ. 156 (2): 151-158.

Mäkelä-Kurtto, R. (1989a) Liukoisen kadmiumin määrä lisääntynyt viljelymaissamme. Koetoim. ja Käyt. 46: 59 (in Finnish).

Page, A.L., Bingham, F.T. and Shang, A.C. (1981) Cadmium. In: Lepp, N.W., ed. Effect of heavy metal pollution on plants, Barking, Essex, Applied Science Publishers, Vol. 1, pp. 77-109.

Schuhmacher, M., Domingo, J.L., Llobet, J.M. and Corbella, J. (1994) Cadmium, chromium, copper, and zinc in rice and rice field soil from southern Catalonia, Spain. Bull. Environ. Contam. Toxicol. 53 (1): 54-60.

Sippola, J. and Erviö, R. (1986) Raskasmetallit maaperässä ja viljelykasveissa Harjavallan tehtaiden ympäristössä. Ympäristö ja Terveys 1986: 270-275.

Tjell, J.C., Hansen, J.A., Christensen, T.H. and Hovmand, M.F. (1981) Prediction of cadmium concentrations in Danish soils. In: L'Hermite, P. and Ott, H., ed. Characterization, treatment and use of sewage sludge, Dordrecht, Reidel Publishing Co., pp. 652-664.

Viitasalo, I. (1978) Heavy metals in soil and cereals fertilized with sewage sludge. Prog. Wat. Tech. 10: 309-316.

Managing Cadmium Contamination of Agricultural Land

M. J. McLaughlin[1,2], K.G. Tiller[1,2] and A. Hamblin[2]

[1]CSIRO Division of Soils

[2]Cooperative Research Centre for Soil and Land Management

Summary

Cadmium concentrations are slowly accumulating in many agricultural soils in Australia. For this reason it is prudent that to protect the soil resource from further contamination some limits be placed on the quality of fertilizers, sewage sludges, limes, gypsums and other soil amendments applied to soils. This is already occurring in most States in Australia.

Cadmium uptake by plants from soil is affected by a number of soil, plant and environmental factors that have varying scope for modification by growers to minimise cadmium concentrations in produce. It has traditionally been accepted that the main method of controlling cadmium accumulation by crops is to increase soil pH by liming. Results from various field trials in Australia on wheat, pastures and potatoes indicate that the benefits of lime in reducing cadmium concentrations in crops are often inconsistent or small in magnitude. Other factors e.g. zinc deficiency, soil salinity or crop cultivar, have been found to have greater influence than soil pH on cadmium accumulation by crops.

To date, the most successful options for minimising cadmium accummulation in Australian crops and livestock appear to be:

1) avoidance of saline soils;

2) avoidance of the use of saline irrigation waters,

3) alleviation of zinc deficiency;

4) minimising weed composition of grazed pastures, especially capeweed;

5) selecting plant species with low cadmium accumulation characteristics;

6) breeding and choosing cultivars with low cadmium concentrations;

7) restricting cadmium accumulator crops to areas with soil and irrigation conditions which minimise cadmium uptake by plants;

8) liming.

Over the last 20 years a large amount of information has been gathered on cadmium concentrations in produce and management factors controlling cadmium concentrations in crops, both from monitoring studies and from research trials. This information indicates that even with good farm management practices on relatively unpolluted soils, some plant species and cultivars regularly exceed the Australian maximum permitted concentrations due to soil, plant and environmental factors.

Introduction

Cadmium contamination of agricultural has become increasingly important in recent years due to increased public awareness and concern for food and land quality. cadmium residues in foods are regularly monitored at both the State and Federal level in Australia and some agricultural produce has been found to exceed maximum permitted concentrations (MPCs) (Stenhouse 1992; Anon 1992a,b). Since the first reports by Schroeder and Balassa (1963) alerting that fertilizers are implicated in elevated cadmium concentrations in food crops, much work has been performed to investigate the impact of cadmium in fertilizers on crop uptake of cadmium. Stringent regulations for cadmium loadings to soils and fertilizers are being considered in some countries of the European Union (Anon 1991; OECD 1994). While the severity of risk associated with dietary intake of certain levels of cadmium is still a contentious issue (Ryan et al., 1982; Friberg et al., 1985), many countries have regulations controlling permissible levels of cadmium in soils and crops. It is now well accepted that concentrations of cadmium in most agricultural soils in Europe, North America and Australia/New Zealand have been elevated due to fertilizer use and industrial activity (Williams and David, 1973; Mordvedt 1987; Merry and Tiller, 1991; Jensen and Bro-Rasmussen, 1992; Nicholson et al., 1994; Roberts et al., 1994).

Given that cadmium concentrations have increased in many improved agricultural soils in Australia, research has recently focussed on soil and environmental factors leading to increased soil-plant transfer of this element and on defining farm management factors which minimise cadmium uptake by crops and intake by animals.

Cadmium inputs to Australian agricultural soils

Tiller et al. (1994) recently reviewed inputs of cadmium to Australian agricultural soils. In contrast to countries in Europe and North America, the major source of cadmium added to agricultural soils in Australia is phosphatic fertilizers. Hence there is a significant relationship between concentrations of phosphorous in soil (from fertilizer applications) and concentrations of cadmium (Figure 1).

Atmospheric accessions of cadmium to most agricultural soils are negligible (Merry and Tiller, 1991) with some notable exceptions in the vicinity of industrial centres e.g. Port Pirie in South Australia (Cartwright et al., 1977). Average annual inputs of cadmium to soils are shown in Table 1.

Table 1

Average annual inputs of cadmium to agricultural soils

Country	Average annual cadmium input g ha^{-1} yr^{-1}	Reference
Australia	1.6	McLaughlin et al. (1995b)
Denmark	3.0	Hovmand (1981)
Germany	3.5-4.3	Kloke et al. (1983)
New Zealand	8.9	Bramley (1990)
UK	4.3	Hutton and Symon (1986)
USA	0.3-1.2	Mordvedt (1987)

Figure 1

Relationship between extractable phosphorous and extractable cadmium in a range of soils under pasture in South Australia

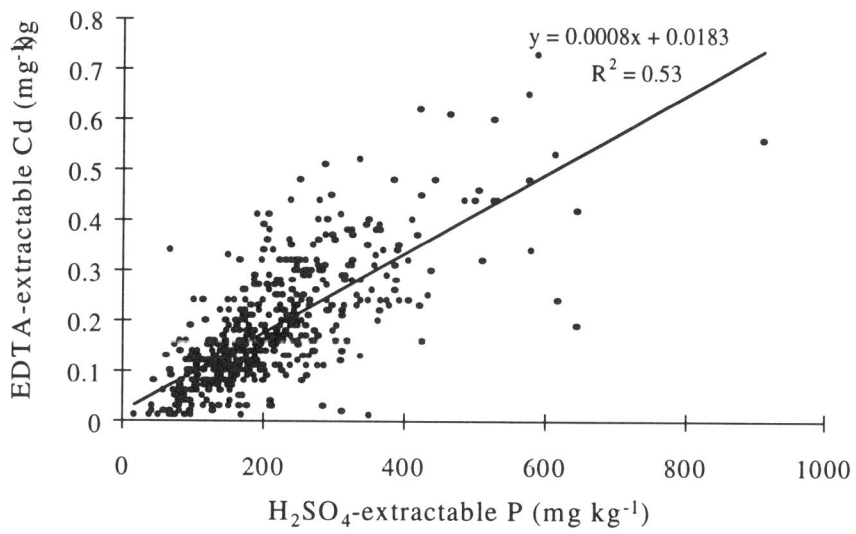

While current inputs of cadmium are low on a per unit area basis when estimated as a national average, certain soils receive higher cadmium loading due to high rates of fertilizer use on some crops e.g. up to 30 g cadmium $ha^{-1}\ y^{-1}$ for potatoes or productive dairy pastures.

Given the current exports of agricultural commodities from Australia, approximately 0.5 tonnes cadmium is exported in produce assuming no cadmium is exported in wool, cotton, milk, dairy products, timber or timber products. We could not source any data for cadmium concentrations in wool, cotton or timber products, but it is likely these commodities would have low cadmium concentrations. Given the current annual import of cadmium in fertilizers and phosphate rocks is approximately 40 tonnes, this mass balance on a national scale confirms that cadmium is slowly accumulating in agricultural soils.

The accumulation of cadmium in soils from long term application of phosphatic fertilizers is now well documented in studies in many other countries (Mulla et al. 1980; Mordvedt et al. 1981; Rothbaum et al. 1986; Baerug and Singh 1990). The major difference between cadmium inputs between Australia and European countries is that fertilizers provide the bulk of the cadmium inputs to agricultural soils, whereas in Europe atmospheric inputs and sewage sludge or waste material sources may be appreciable. Many reports quote approximately equal inputs of cadmium from atmospheric and fertilizer sources, a situation rarely applicable in Australia except near major urban centres. Furthermore, most Australian soils are naturally low in cadmium whereas some soils in Europe and North America have naturally high cadmium concentrations.

Sewage sludge, until recently, has not been applied in appreciable amounts or to significant areas of Australian agricultural land. This situation will soon change because of growing government and community reaction to incineration or ocean disposal of sewage, treated or otherwise. Tiller et al. (1994) have recently estimated potential sludge-cadmium loading to Australian agricultural soils to be about 2-4 tonnes y^{-1}.

Plant uptake of cadmium from soil

It is generally thought that the chemical form of cadmium taken up by plants is the free uncomplexed Cd^{2+} ion present in soil solution (for reviews see Chaney and Hornick, 1988 and Parker et al. 1994). Thus, any treatments or changes in soil conditions which affect the concentration (or activity) of the Cd^{2+} ion will affect plant accumulation of cadmium. The soil and plant factors controlling cadmium uptake from soils have been extensively studied and reviewed (Chaney and Hornick 1978; Page et al. 1987; Jackson and Alloway 1992). Table 2 summarises the factors listed by Chaney and Hornick (1978) as affecting plant uptake of cadmium from soil.

Most of the above reviews focus on the cadmium contamination of soils by addition of sewage sludges to land, with assessment of subsequent plant uptake of added cadmium. Although many of the soil factors discussed also apply to the behaviour of fertilizer-derived cadmium, the amounts and form of cadmium added in sewage sludge may lead to very different reactions in the soil which affect plant uptake of metals. This is an area still being actively researched in relation to the policy of disposal of sludge to land.

Table 2

Factors affecting cadmium uptake by plants from soil
(adapted from Chaney and Hornick, 1978)

Factors	Effects on cadmium uptake by plants
Soil	
1. pH	uptake increases as pH decreases
2. soil salinity	uptake increases as salinity increases
3. amount of cadmium present	uptake increases with concentration increases
4. metal sorption by soil	uptake decreases as sorption increases
a. organic matter	higher org. matter generally decreases uptake
b. cation exchange capacity (CEC)	high CEC reduces uptake
c. clay, Fe and Mn oxides	presence decreases uptake
5. micronutrients	e.g. zinc deficiency increases uptake
6. macronutrients: NH4, PO4, K	may increase or decrease uptake
7. temperature	higher temperatures increase uptake
8. aeration	e.g. flooding reduces uptake
Crop	
1. species and cultivar	leafy vegs.> root vegs.>cereals>fruits
2. plant tissue	leaf >grain, fruit and edible root.
3. leaf age	older > younger
4. metal interactions	presence of zinc reduces uptake of cadmium

Management of soil factors affecting cadmium uptake by plants

1. Changing soil acidity (pH)

Soil pH is often regarded as the major variable controlling plant uptake of cadmium from soils (Chaney and Hornick 1978), with cadmium uptake generally decreasing as soil pH increases. Data on effects of soil pH on cadmium uptake under field conditions have been reported from surveys of sludge-contaminated soils (Alloway et al. 1990) or normal agricultural soils (Sillanpaa and Jansson 1992; He and Singh 1993) where plant cadmium concentrations were related to field soil properties and acidic soils were found to have plants with the highest cadmium concentrations. Results from many glasshouse and laboratory liming experiments have indicated that soil pH has a major effect on cadmium uptake by plants (CAST 1976; Miller et al. 1976; Williams and David 1976; Jackson and Alloway 1991).

Thus liming of agricultural soils has been identified as a management tool to reduce cadmium uptake by crops. However, the effectiveness of reducing crop cadmium concentrations with lime under field conditions is not clear cut. Some authors report significant reductions in crop cadmium concentrations with lime under field conditions (CAST 1976; CAST 1980; Mordvedt et al. 1981) but compared to glasshouse trials the differences noted due to liming under field conditions are inconsistent and of a smaller magnitude (Merry, 1992; Tiller et al. 1994; Oliver et al. 1994a; Maier and McLaughlin, unpublished data). In other studies, lime had little or no effect on crop cadmium concentrations in both sludge-amended soils (Pepper et al. 1983) and fertilized agricultural soils (Jaakkola 1977; Sparrow et al. 1993a) and in some sites may even increase crop cadmium concentrations (Andersson and Siman 1991; Sparrow et al. 1993a; Maier and McLaughlin, unpublished data - Figure 2).

These conflicting results are of concern as liming is traditionally regarded as the main management tool available to reduce cadmium accumulation by crops. Further research is required to determine under which conditions lime may or may not be effective.

2. Changing or regulating soil cadmium concentrations

The total cadmium concentration in soil has been found to be of equal importance to soil pH in many studies examining plant cadmium uptake (Chumbley and Unwin 1982; Jackson and Alloway 1991; Sillanpaa and Jansson 1992; He and Singh 1993). Generally, the higher the cadmium concentration in soil the greater the plant cadmium concentration. Cadmium concentrations in soil are determined by the composition of the parent rock material and the historical land use.

Changing soil concentrations is virtually impossible under field conditions, but the rate of accumulation of cadmium in agricultural soils can be controlled through regulation of the quality of fertilizer, lime (and other soil amendments) and sewage sludges used on agricultural soils.

Figure 2

Effect of liming on cadmium concentrations in potato tubers in southern Australia

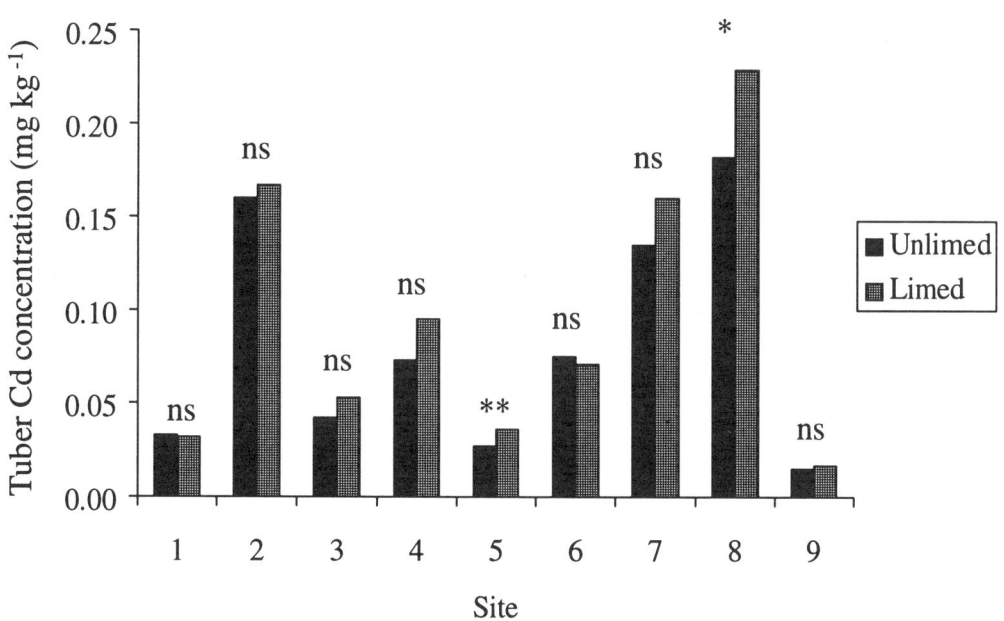

Standards for contaminants in fertilizers are currently under review in Australia with most States either having introduced legislation or having legislation pending. Tasmania and Victoria have set a maximum permitted concentration for cadmium in fertilizers. Other States have legislation pending. Cadmium limits are shown in Table 3 (Anon, 1993a).

Regulations controlling disposal of sludge to agricultural land are shown in Table 4.

However, changing the quality of fertilizers applied to agricultural land may not have any immediate impact on crop cadmium concentrations, due to the contribution of cadmium already present in the soil to crop uptake. For example, McLaughlin et al. (1993) found that switching to low cadmium fertilizers for potato production had little effect on cadmium concentrations in the crop immediately following the switch to low cadmium products (Figure 3).

Table 3

Current existing or proposed legislation regarding cadmium contents of fertilizers in Australia and in Australia's trading partners (from McLaughlin et al. 1995b)

Country	Cadmium Level (mg cadmium kg^{-1} P*)
Austria	275
Denmark	150 from July 1992
	110 from July 1995
France	**None**
Germany	200 voluntary
Japan	340
Netherlands	40
Norway	100
Sweden	over 5 = tax of 30SEK mg^{-1} cadmium
	100 maximum
Switzerland	50 from 1992
UK	**None**
USA	**None**
Australia	
Tasmania	
Phosphatic fertilizers	450
Trace element fertilizers	80 (per kg product)
Soil amendments	10 (per kg product)
Victoria (proposed)	
Phosphatic fertilizers	350
Other fertilizers	10 (per kg product)

* *the phosphorus concentration of the fertilizer is the unit of significance against which cadmium concentration should be measured, rather than cadmium concentrations per unit weight of rock.*

Table 4

Current existing or proposed legislation regarding maximum permissible cadmium concentrations in sludges applied to agricultural land (OECD, 1994)

Country	Sludge cadmium concentrations (mg cadmium kg^{-1})	Limit on cadmium loading to agricultural land (g ha^{-1})
Australia (NSW)	3	-
Austria	5	25 y^{-1} (crops) 12.5 y^{-1} (pasture)
Canada	20	4000 over 45 y
Finland	-	3 y^{-1}
Germany	5-10 depending on soil	-
Netherlands	1.25	-
Norway	10	-
New Zealand		200 y^{-1}
Sweden	4	-
Switzerland	5	-
UK	-	150 over 10 years
USA	39	500

Figure 3

Effect of fertilizer type and quality on concentrations of cadmium in potato tubers at three sites in southern Australia (from McLaughlin et al., 1993)

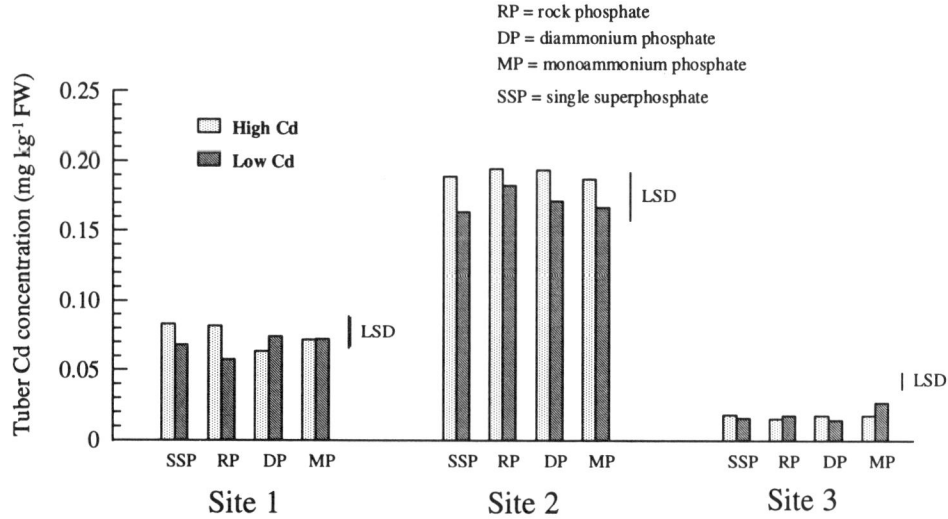

Regulations controlling cadmium concentrations in soils are driven more from consideration of gross contamination of soils from urban or industrial sources, rather than the low annual loading rates due to fertilization of agricultural soils. These limits often relate to Dutch soil standards which were set in the mid 1980s (Moen et al. 1986). These criteria are applied in relation to the health risk associated with the ingestion of contaminated soil by children. Environmental investigation guidelines have yet to be established in relation to phytotoxic effects, production of foods of unacceptable quality or perturbations to the natural environment. Some suggested soil quality guidelines have been proposed by NHMRC/ANZECC (1992). Table 5 summarizes some soil background values and several proposed levels for cadmium at which some investigation of the nature of the contamination must be made and health risk assessments performed.

Table 5

Background metal concentrations (mg kg^{-1}) and investigation levels for soils in various countries

	Background Values			Investigation Levels		
	Australia Tiller (1991)	Australia - ANZECC/ NHMRC (1992)	USA Holmgren et al. (1993)[a]	Dutch A Value	ANZECC/ NHMRC (1992)	Dutch B Value
Cadmium	<1	0.04-2	<0.01-2.0 (0.26)	1	3	5

[a] *Means USA values shown in brackets.*

3. Bioremediation

It has been proposed that hyperaccumulator plant species may be used to reduce toxic metal concentrations in soils through harvest of plant materials with high metal concentrations. This option is purely experimental and even if proven technically possible, the economic and logistical problems of decontaminating broad-acre agricultural soils must make this an unlikely option for cadmium management in broadacre agriculture. The technology may have some application in urban or industrial situations where high land values allow more sophisticated techniques to be applied.

4. Reducing plant uptake of cadmium by changing affinity of soil for cadmium

The control of the activity of Cd^{2+} ion in soil solution can be used to reduce cadmium uptake by plants. High affinities (strong sorption) of cadmium by soil particles will tend to reduce the activity of Cd^{2+} in the soil solution (after cadmium dissolves from the fertilizer granule or sludge material) and therefore reduces plant availability of inorganic cadmium added to soil (Miller et al. 1976; Chaney and Hornick 1978; Hinesly et al. 1982). Unfortunately, practical large scale options to change the sorption capacity of soil for cadmium (apart from liming) are limited. Experimental approaches to amend soils with highly sorbing materials (e.g. zeolites, iron oxides/hydroxides) have been reported in the literature (Mensch and Martin, 1991), but the efficacy and economics of this strategy under broad-acre field conditions remain unproven, with the economics for Australian agricultural soils being questionable given current commodity prices.

5. Altering soil micronutrient status

Another important factor controlling cadmium accumulation by plants is the micronutrient status of soils, particularly zinc. Chaney and Hornick (1978) reviewed early data on cadmium:zinc interactions and recently Tiller et al. (1994) reviewed more recent data. Most studies of effects of zinc on cadmium uptake have been in glasshouse experiments, often at unnaturally high concentrations of both cadmium and zinc (John et al. 1972; MacLean 1976; Williams and David 1976; White and Chaney 1980) and results have not been conclusive (Tiller et al. 1994). In Australia, Oliver et al. (1994b) recently demonstrated that where soils are zinc deficient, small amounts of zinc (<10 kg zinc ha-1 which is easily applied with normal fertilizer applications) significantly reduced cadmium concentrations in wheat grain (Figure 4). It is unknown the extent to which zinc deficiency is responsible for occurrences of localised high cadmium concentrations in some produce, but zinc deficient soils are widespread in many areas of SA, WA and Queensland.

Even on zinc adequate soils (Figure 5) experiments with potatoes have shown that significant reductions in tuber cadmium concentrations can be achieved with addition of zinc (McLaughlin et al. 1993), but the magnitude of the effect was smaller (c.10-15 per cent reduction) than those noted by Oliver et al. (1994b).

More recent data of Oliver et al. (unpublished) indicate that additon of zinc in excess of nutritional requirements continues to depress cadmium uptake by wheat (Figure 6).

Figure 4

Effect of small applications of zinc on grain cadmium concentrations in wheat at two sites in South Australia (from Oliver et al., 1994b)

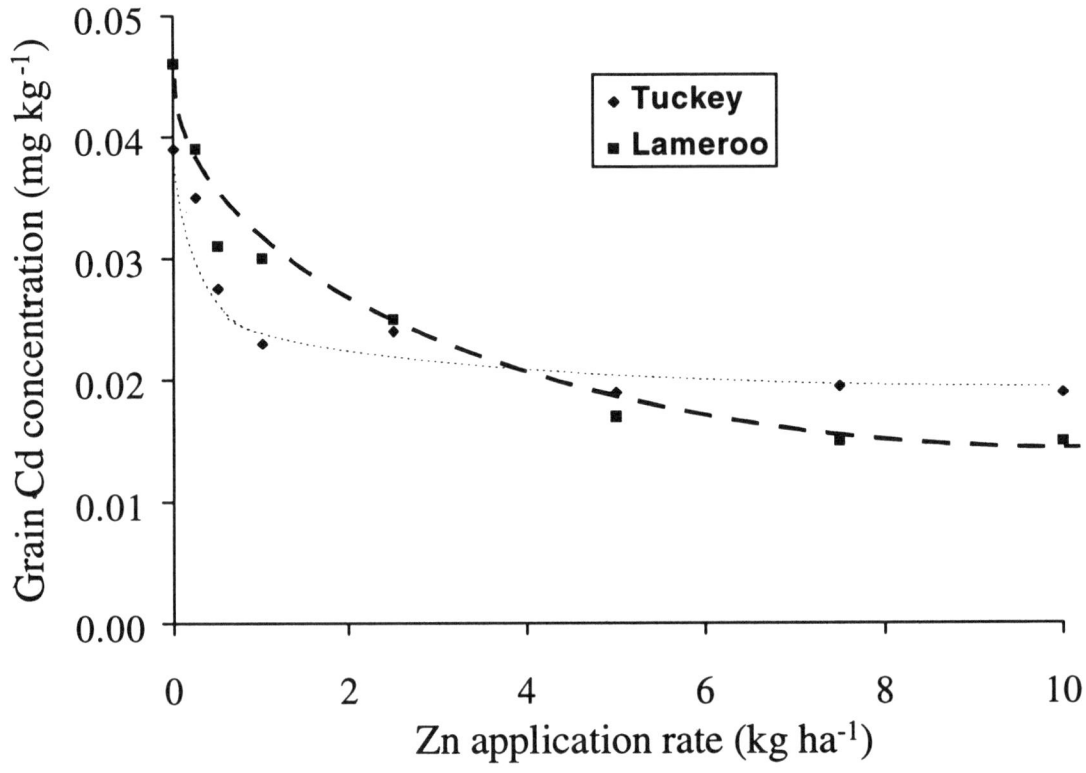

Figure 5

Effect of increasing rate of zinc (as zinc SO_4) applied at planting on concentrations of cadmium in potato tubers (from McLaughlin et al. 1993)

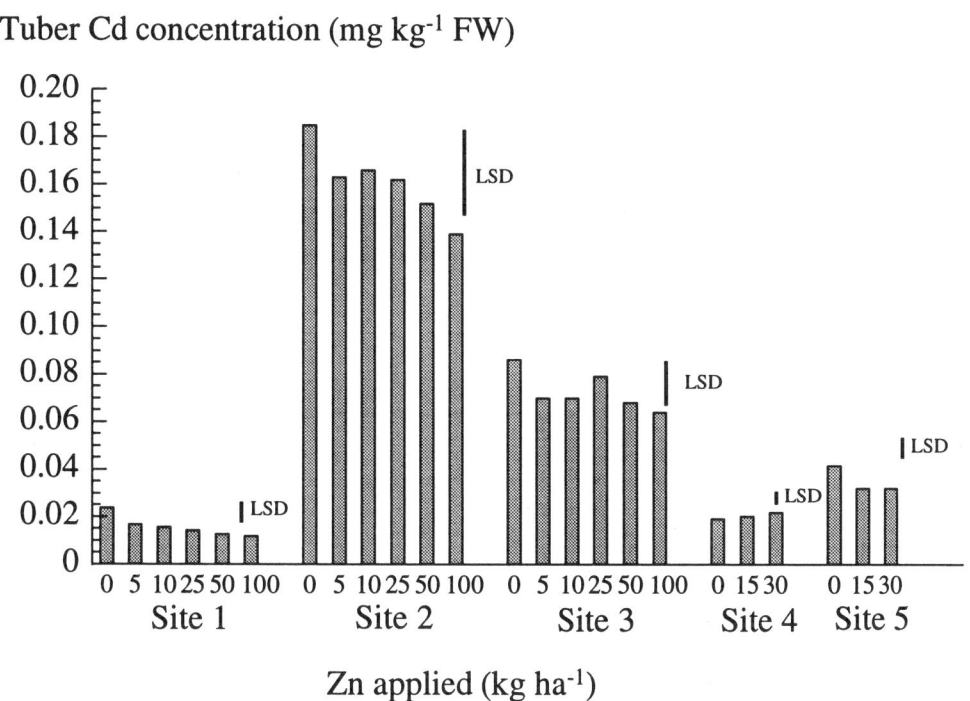

Figure 6

Effect of high applications of zinc on grain cadmium concentrations in wheat at two sites in South Australia (from Oliver et al., unpublished data)

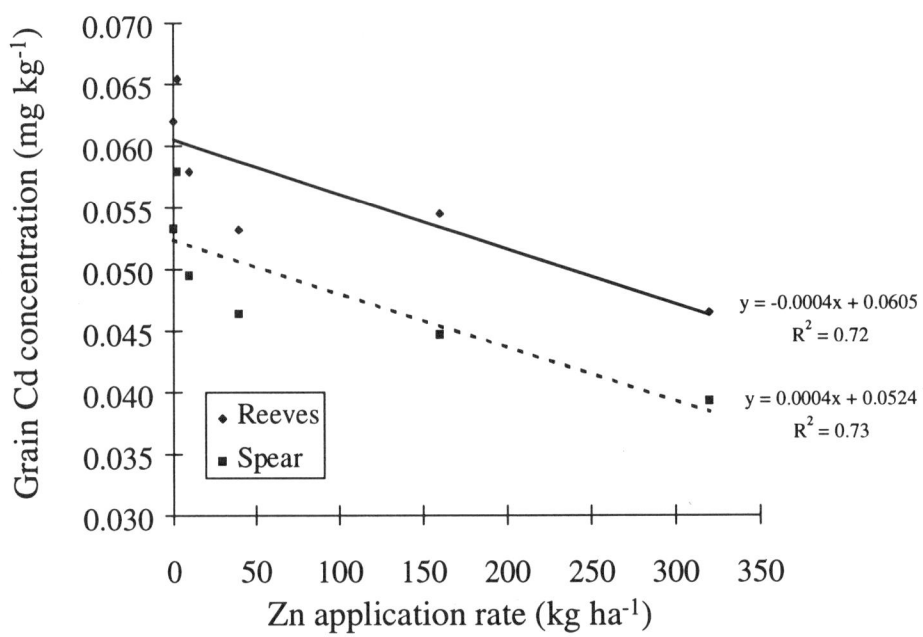

Such high application rates for zinc (>250 kg ha^{-1}) may be difficult to justify on economic grounds, but alleviation of zinc deficiency by applications of small amounts of zinc is certainly a promising management tool to reduce cadmium accumulation in crops, as well as significantly improving crop productivity where soils are zinc deficient.

6. Effect of changing soil salinity

Recent research results from our laboratory have indicated that soil salinity is a major factor controlling cadmium accumulation by potato crops in Australia (McLaughlin et al. 1994a). A range of commercial potato crops were sampled with the aim of determining the soil factors responsible for regionally high tuber cadmium concentrations. The results indicated that soil salinity was the major factor associated with high tuber cadmium concentrations (Figure 7).

Figure 7

Relationship between tuber cadmium concentrations and extractable Cl concentrations in soil (from McLaughlin et al., 1994)

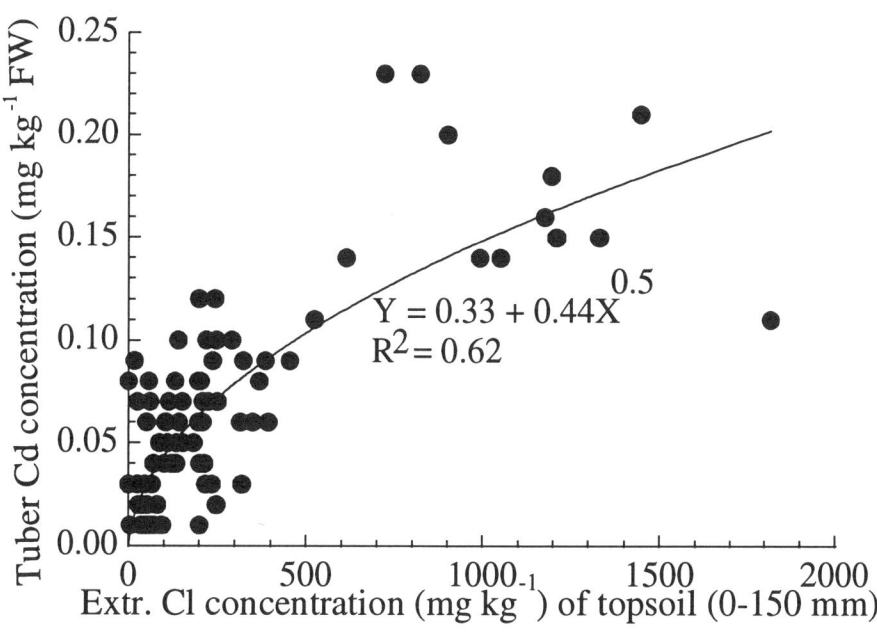

Soil salinisation can be caused by either additions of chloride or sulfate salts in irrigation waters, or by the rise to the surface of these salts in groundwaters (dryland salinisation). Data from our laboratory indicate that the effect appears to be limited to chloride concentrations in soil and not sulfate. As soil salinisation is a major and increasing problem in many regions of Australia, it is likely that this is a significant issue for crop cadmium concentrations in future. Recent data from our laboratories indicate that similar effects of salinity on cadmium uptake are evident for pasture legumes (McLaughlin et al., unpublished data). A major concern with salinisation is that crops with high cadmium concentrations may be produced on soils having relatively low cadmium concentrations. In dryland or irrigation areas where water tables are rising, cadmium in the topsoil will be mobilised as salts slowly begin to accumulate at the surface.

Controlling dryland salinisation is a long-term issue for soil conservation irrespective of cadmium issues. Similarly, control of salinity in irrigation regions is a major focus of most water management systems in the irrigation areas. If cadmium is a health or marketing issue and high value crops which are likely to accumulate cadmium must be grown, the only immediate remedy is to locate production in areas where the soils are non-saline and/or irrigation waters are of a high quality.

7. Altering the macronutrient status of soils

Other factors identified as affecting plant uptake of cadmium from soils have been soil temperature (Giordano et al. 1979; Hooda and Alloway 1993), reducing conditions (Bingham et al. 1976) and addition of macronutrients phosphorous and N (John et al. 1972; Miller et al. 1976; MacLean 1976; Williams and David 1976, 1977; Jaakkola 1977; Eriksson 1990; Willaert and Verloo 1992). Soil temperature and soil redox potential are factors over which there is little control and the results have little relevance to management of cadmium in broadacre agriculture in Australia. They will not be discussed further here.

Effect of nitrogen fertilizers

Additions of N and phosphorous to soils, in terms of amounts and chemical form, may be altered by farm management practices. For example, Williams and David (1976) found that wheat crops fertilized with superphosphate and ammonium nitrate accumulated more cadmium than crops fertilized with superphosphate only (Figure 8), but were unable to suggest the reason for this effect.

Figure 8

Effects of nitrogen fertilizer and superphosphate on cadmium concentrations in wheat grain

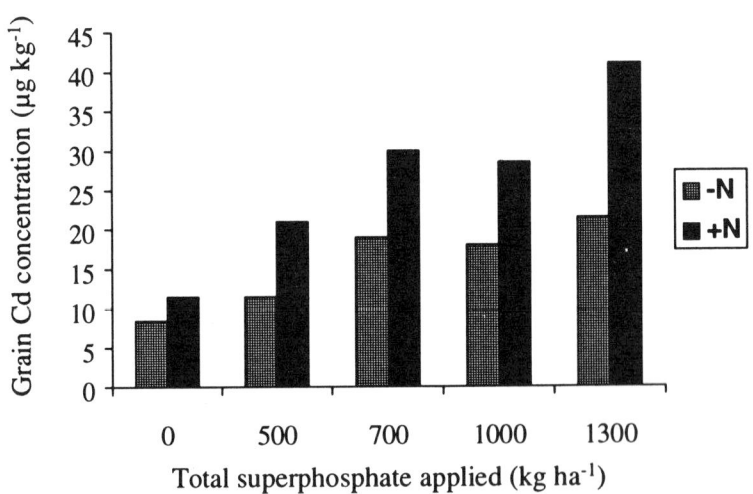

(from Williams and David, 1976)

All ammonium-based fertilizers induce localised acidification in soil (close to the fertilizer granule), which with repeated and high applications will lead to acidification of the whole soil profile. Indeed, Eriksson (1990) and Willaert and Verloo (1992) both demonstrated in glasshouse experiments that more acidifying N fertilizers, such as ammonium sulfate, increased cadmium uptake over those having an alkaline effect in soil (nitrate-based). Similar results were obtained by Reuss et al. (1978). However, there are no data from field experiments which support these findings. Indeed, McLaughlin et al. (1993) found no effect of changing phosphorous fertilizer type (banded) on cadmium uptake by potatoes under field conditions (Figure 9).

Effect of potassium fertilizers

Potassium sulfate and potassium chloride are the two major forms of potassium added to soil in fertilizers. Potassium itself has little effect on cadmium availability, but the associated sulfate or chloride may influence cadmium uptake by plants. Sparrow et al., (1994) found that in four out of six sites potato crops fertilized with potassium sulfate instead of potassium chloride had slightly lower tuber cadmium concentrations. However, in previous work McLaughlin et al. (1993) found no effect on plant cadmium concentrations of changing potassium fertilizer type. Possible reasons for these conflicting results are that the rates of potassium fertilizer (and hence Cl) applied by Sparrow et al. (1994) were higher (160-350 kg potassium ha^{-1}) than those used by McLaughlin et al. (1993) (150 kg potassium ha^{-1}) and irrigation water quality was better in the studies of Sparrow et al. (1994).

Effect of adding phosphorous and phosphorous placement

The effect on plant cadmium uptake of adding phosphorous (without cadmium) to soils has also been studied, as well as placement of phosphorous in the root zone in both glasshouse (Williams and David 1977; Reuss et al. 1978) and field conditions (Sparrow et al. 1992, 1993b). In glasshouse experiments, Williams and David (1977) noted that adding phosphorous to a soil increases cadmium uptake through a stimulation of root proliferation in the zone into which phosphorous is added. Sparrow et al. (1993b) compared cadmium uptake by potatoes fertilized with both low- and high-cadmium DAP in field trials and found little differences in cadmium uptake between the two sources, with cadmium concentration in tubers being related to the rate of phosphorous applied, rather than the rate of cadmium applied. These data support the early results of Williams and David.

Figure 9

Effect of changing phosphorous and potassium fertilizer type on cadmium concentrations in potato tubers at three sites in southern Australia (from McLaughlin et al., 1993)

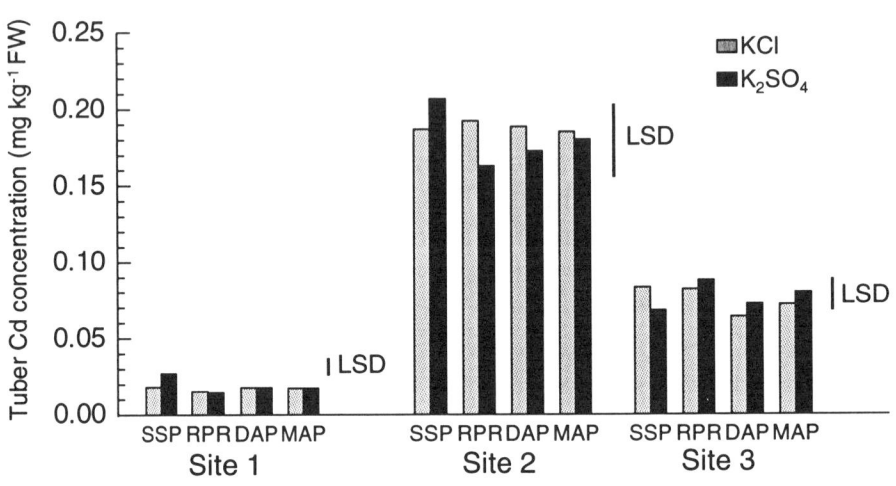

8. Changing crop rotations

In Australia cereal crops are often grown in sequence with other crops such as legumes, e.g. peas, beans, lupins, and clover- or medic-based pastures. Such rotations have been a common management practice to minimize root disease, control weeds and increase nitrogen status. Lupins are being increasingly introduced as a break crop in cereal production areas. However, negative impacts such as some soil acidification are likely in poorly buffered soils, especially crops such as lupins that are known to acidify soil.

The effects of various crop rotations on cadmium uptake by wheat were investigated at two field trial sites, Kapunda and Tarlee in South Australia (Oliver et al., 1993). Data from the Tarlee trial are shown in Figure 10. Legume crops in the rotation, and especially lupins, increased cadmium concentrations in wheat grain. Results from the Kapunda site were similar and recent data from trials in WA have confirmed these results.

Figure 10

Effect of crop rotations on cadmium concentrations in wheat at Tarlee, South Australia (from Oliver et al., 1993)

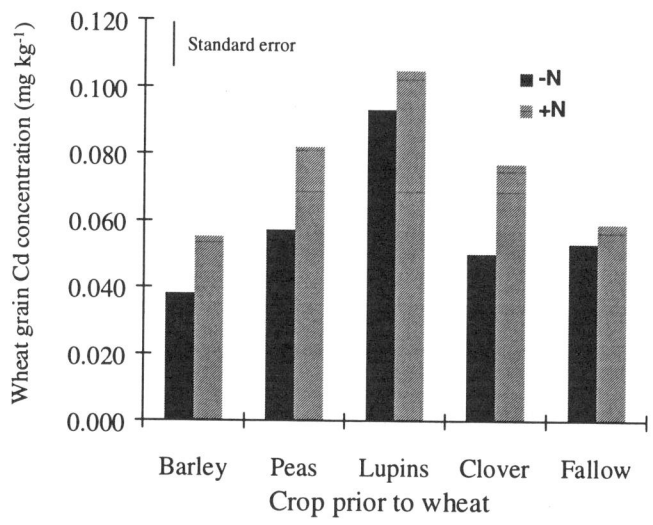

It is not possible yet to fully explain the increased cadmium concentrations in wheat grain following lupins. The effect may be related to a decrease in soil pH commonly observed with the growth of legumes or to decomposition products of the lupin stubble or roots in the soil. These aspects are currently under investigation.

Lupins are an important component of sustainable agricultural systems in southern Australia and their inclusion in crop rotations is favoured by growers due to benefits due to disease minimisation and N-fixing benefits. Sound management strategies are needed to maximize their benefit to agricultural production yet minimise adverse impacts on marketability of grain.

9. Changing tillage and stubble handling practices

There are few data to indicate the impact of changing tillage or stubble handling on grain cadmium concentrations in cereals. At the Kapunda site in SA, cadmium concentrations of direct drilled wheat in 1989 were about one third higher than for either conventional or reduced tillage (Oliver et al., 1993). This was postulated to occur due to a restriction of root growth to the surface soil horizon where cadmium is concentrated. The lack of significant effects of cultivation on grain cadmium in the subsequent year indicate the seasonality of this factor.

Different methods of treating wheat stubble were also investigated at the Tarlee crop rotation trial because of their possible effect on soil organic matter status and thus indirectly cadmium uptake by a later wheat crop. Treatments included burning, incorporation into the cultivated soil and retention of straw on the soil surface. None of these treatments had a significant effect on grain cadmium concentration in the two years tested.

10. Changing pasture composition

Differences in cadmium concentrations between species are notable in pastures (Figure 11), where capeweed (*Arctotheca calendula*) has been found to have much higher cadmium concentrations than either medics (*Medicago* sp.) or ryegrass (*Lolium* sp.) (Merry and Tiller, 1991). Capeweed is a common component of pastures in many regions of south-eastern and south-western Australia and provides useful summer forage for grazing animals in some areas. In constructing a budget for cadmium intake by sheep, Merry (1992) indicated that the prevalence of capeweed in pastures is the single most important factor determining cadmium uptake by grazing animals.

Hence pasture management to control capeweed growth will have a major impact on cadmium uptake by stock.

Figure 11

Concentrations of cadmium in pasture species and grain legumes grown at Tarlee, South Australia (from Merry, 1992)

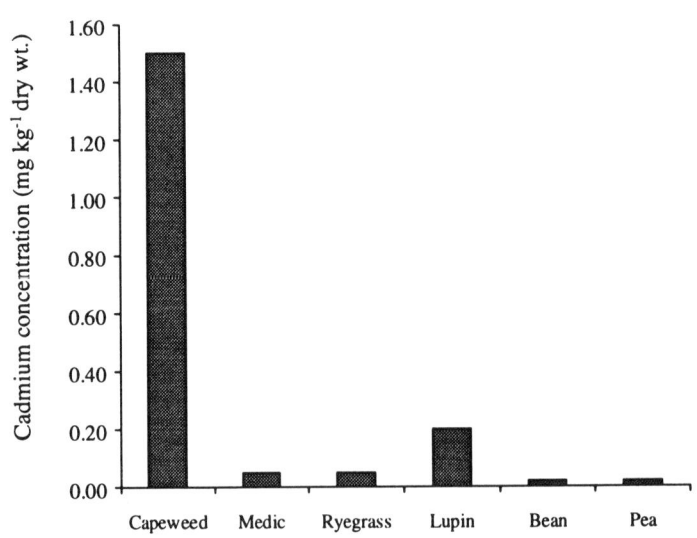

11. Changing plant cultivar

Cultivar differences in cadmium uptake have been reported for a number of food crops. Australian data have been limited to wheat (Oliver et al., 1995) and potatoes (McLaughlin et al., 1994b). While we have found in field screening studies that site and soil factors play a dominant role in determining absolute crop cadmium concentrations, the relative differences between current commercial cultivars are often significant enough to warrant including cadmium as a selection factor in breeding programs (Figure 12) (McLaughlin et al. 1994b; Oliver et al., 1995). Larger variation may be found when advanced breeding lines are compared, so there appears to be scope to select for low cadmium accumulating genotypes.

Figure 12

Mean cadmium concentrations for major potato cultivars averaged over a range of sites throughout Australia: n=4 to 8 for any single cultivar (from McLaughlin et al., 1994b)

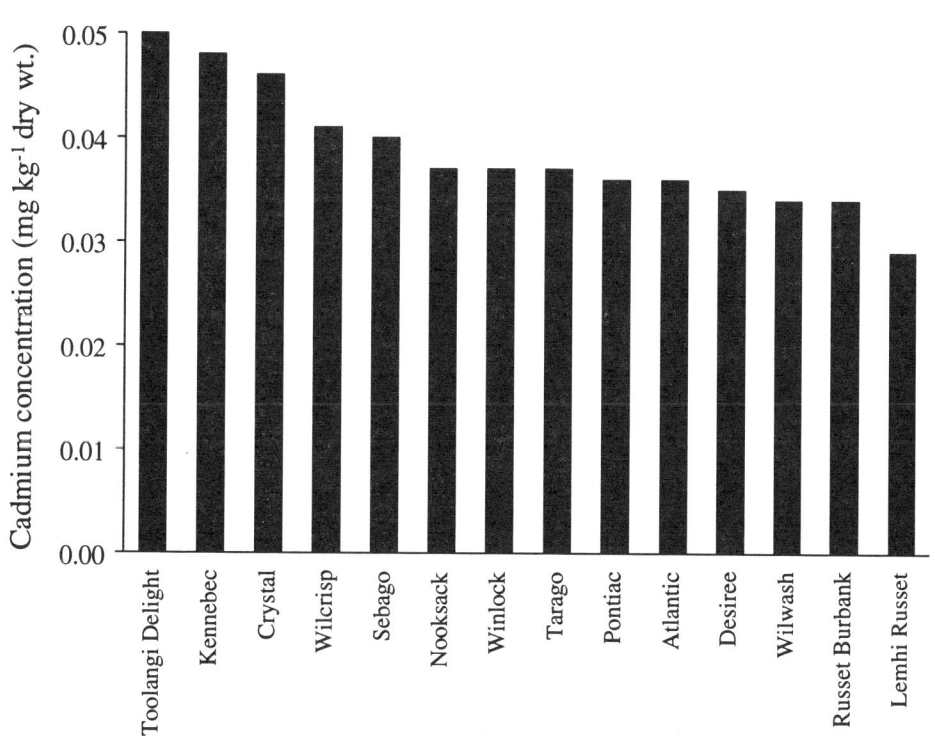

Cadmium concentrations in crops in Australia

Cadmium concentrations in produce in other countries are generally similar to concentrations in Australian produce (Table 6).

While Australian soils are generally less polluted and generally have lower background cadmium concentrations than soils in North America and Europe, our soils are generally lighter textured, more acidic (through lack of lime use), more likely to be zinc deficient and more saline than soils in Europe or North America. All these factors tend to enhance the transfer of cadmium from soil to crops so that relatively high cadmium concentrations can be found in crops growing on soils with relatively low total or available cadmium concentrations. This point indicates that attempting to control entry of cadmium into the food chain will not be easily and quickly accomplished by regulating permissible cadmium concentrations in soil or fertilizers. For example, as a consequence of soil salinity, potato tubers having up to 0.25 mg cadmium kg^{-1} (fresh weight) can be produced on soils with total cadmium concentrations less than 0.2 mg kg^{-1}. However, it is prudent that entry of further cadmium into agricultural soils be minimised so that regulation of the quality of fertilizers, sludges, limes and other soil amendments is necessary.

Conclusions

While Australian soils have relatively low cadmium concentrations by international standards, cadmium concentrations in foodstuffs are similar to concentrations in other countries. This may be due to the fact that Australian soils are generally lighter textured, more acidic, more saline and more likely to be zinc deficient (all of which enhance plant uptake of cadmium) compared to soils in North America and Europe.

The accepted method for controlling cadmium accumulation in crops is liming, but the results in Australia are often inconsistent and although liming is necessary to ameliorate soil acidity, it cannot be relied upon alone to control cadmium accumulations in crops. Soil pH undoubtedly influences cadmium availability but it has yet to be determined under which conditions the application of lime will substantially reduce crop cadmium concentrations. To date, the most successful options for minimising cadmium accummulation in crops appear to be (in no priority):

1) alleviation of zinc deficiency;

2) avoidance of saline soils;

3) avoidance of the use of saline irrigation waters;

4) selecting plant species with low cadmium accumulation characteristics;

5) minimising weed composition of grazed pastures, especially capeweed;

6) breeding and choosing cultivars with low cadmium concentrations;

7) restricting cadmium accumulator crops to areas with soil and irrigation conditions which minimise cadmium uptake by plants;

8) liming.

Over the last 20 years a large amount of information has been gathered on cadmium concentrations in produce and management factors controlling cadmium concentrations in crops, both from monitoring studies and from research trials. This information indicates that even with good farm management practices on relatively unpolluted soils, some plant species and cultivars regularly exceed the Australian maximum permitted concentrations due to soil, plant and environmental factors.

Table 6

Cadmium concentrations in some agricultural produce

Crop/ Country	No. samples	Mean	Median	Range	Reference
Wheat					
Australia - Queensland	267	0.006	NR	<0.001-0.042	Best (1991).
Australia - SA	1,751	0.015	NR	<0.002-0.256	Richards et al. (pers.comm.).
USA - All States	288	0.043	0.030	<0.002-0.207	Wolnik et al. (1983).
Germany	886	0.056	0.046	0.008-0.849	Weigert et al. (1984).
United Kingdom	20	<0.050	NR	<0.010-0.090	Anon (1983).
Barley					
Australia - all states	83	0.002-0.009a	NR	0.001-0.018	Anon (1990).
Australia - SA	1,557	0.010	NR	<0.002-0.073	Richards et al. (pers.comm.).
United Kingdom	21	<0.030	NR	<0.010-0.050	Anon (1983).
Potatoes					
Australia - all States	80	0.048	0.040	NR-0.140	NRS (pers. comm.).
Australia - all States	32	0.040	0.030	NR-0.130	Stenhouse (1991).
Australia - all States	429	0.037	0.046	0.004-0.232	McLaughlin et al. (unpubl. data)
Australia - WA	116	0.035	0.030	0.005-0.120	Anon (1992c).
Netherlands	94	0.030	0.030	0.010-0.080	Wiersma et al. (1986)
USA - all States	297	0.031	0.028	0.002-0.181	Wolnik et al. (1983).
Germany	133	0.047	0.030	0-0.320	Weigert et al. (1984).
Onions					
Australia - all States	24	0.040	0.020	NR-0.170	Stenhouse (1991).
United Kingdom	11	0.040	NR	0.010-0.090	Thomas et al. (1972).
USA - all States	230	0.011	0.009	0.001-0.054	Wolnik et al. (1985).

a State means

References

Alloway, B. J., Jackson, A. P. and Morgan, H. (1990). The accumulation of cadmium by vegetable grown on soils contaminated from a variety of sources. Sci. Tot. Environ. 91, 223-236.

Andersson, A. and Siman, G. (1991). Levels of cadmium and some other trace elements in soils and crops as influenced by lime and fertilizer level. Acta Agric. Scand. 41, 3-11.

Anon (1983). "Survey of cadmium in food: First supplementary report." Ministry of Agriculture, Fisheries and Food Surveillance Paper No.12. (HMSO, London.).

Anon (1990). Cadmium and lead in Australian barley. Final Report for Grains Research and Development Corporation project "Screening of barley for cadmium and lead content at selected depots within Australia." (Queensland Department of Primary Industries, Indooropilly, Queensland).

Anon (1991). Policy on cadmium. Ministerie van Volkshuisvesting, Ruimtelijk Ordening en Milieubeheer, Directoraat-Generaal Milieubeheer, The Netherlands. 29pp.

Anon (1992a). Report on the national residue survey, 1989-1990 Results. (Bureau of Rural Resources, Australian Government Printing Office: Canberra).

Anon (1992b). Victorian produce monitoring: Results of residue testing 1991-1992. Victorian Department of Food and Agriculture Research Report Series No.105 (September 1992).

Anon (1992c). Cadmium in Western Australian potatoes. Chemistry Centre (WA)/West Australian Department of Agriculture/Health department of Western Australia, Perth.

Anon (1993a). Food standards code. National Food Authority, Australian Government Printing Service, Canberra.

Baerug, R. and Singh, B. R. (1990). Cadmium levels in soils and crops after long-term use of commercial fertilizers. Norweg. J. Agric. Sci. 4, 251-260.

Best, E. K. (1991). Cadmium and lead in Queensland wheat. Final Report for Grains Research and Development Corporation project "Screening of wheat for cadmium and lead content." (Queensland Department of Primary Industries, Indooropilly: Queensland).

Bingham, F. T., Page, A. L., Mahler, R. J. and Ganje, T. J. (1976). Cadmium availability to rice in sludge-amended soil under "flood" and "nonflood" culture. Soil Sci. Soc. Am. J. 40, 715-719.

Bramley, R.G.V. (1990). Cadmium in New Zealand agriculture. N. Z. J. Agric. Res. 33, 505-519.

Cartwright, B.C., Merry, R.H. and Tiller, K.G. (1977). Heavy metal contamination of soils around a lead smelter at Port Pirie, South Australia. Aust. J. Soil Res. 15, 69-81.

CAST (1976). Application of sewage sludge to cropland: appraisal of potential hazards of the heavy metals to plants and animals. Report No. 64, Council for Agricultural Science and Technology. (CAST: Ames, Iowa).

CAST (1980). Effects of sewage sludge on the cadmium and zinc content of crops. Report No. 83, Council for Agricultural Science and Technology. (CAST: Ames, Iowa.)

Chaney, R. L. and Hornick, S.B. (1978). Accumulation and effects of cadmium on crops. In "Proceedings of the First International Cadmium Conference, San Francsico" pp. 125-140. (Metals Bulletin Ltd.: London).

Chumbley, C. G. and Unwin, R. J. (1982). Cadmium and lead content of vegetable crops grown on land with a history of sewage sludge application. Environ. Pollut. 4, 231-237.

Eriksson, J. E. (1990). Effects of nitrogen-containing fertilizers on solubility and plant uptake of cadmium. Water Air and Soil Poll. 49, 355-368.

Friberg, L., Elinder C.G., Kjellstrom T. andNordberg G.F. (1985). Cadmium and health: a toxicological and epidemiological appraisal. Effects and response. CRC Press, Boca Raton, Florida.

Giordano, P. M., Mays, D. A. and Behel, A. D. Jnr. (1979). Soil temperature effects on uptake of cadmium and zinc by vegetables grown on sludge-amended soil. J. Environ. Qual. 8, 233-236.

He, Q. B. and Singh, B. R. (1993). Plant availability of cadmium in soils. II. Factors related to the extractability and plant uptake of cadmium in cultivated soils. Acta Agric. Scand. 43, 143-150.

Hinesly, T. D., Redborg, K. E., Ziegler, E. L. and Alexander, J. D. (1982). Effect of soil cation exchange capacity on the uptake of cadmium by corn. Soil Sci. Soc. Am. J. 46, 490-497

Holmgren, G. G. S., Meyer, M. W., Chaney, R. L. and Daniels, R. B. (1993). Cadmium, lead, zinc, copper, and nickel in agricultural soils of the United States of America. J. Environ. Qual. 22, 335-348.

Hofmand, M. F. (1981). Cirkulation af bly, cadmium, kobber, zink og nickl i dansk landbrug. In "Slammets jordbrugsandvendels, Vol. II Fokusering" pp. 85-118. (Polyteknisk Forlag: Lyngby).

Hooda P. S. and Alloway, B. J. (1993). Effects of time and temperature on the bioavailability of cadmium and Pb from sludge-amended soils. J. Soil Sci. 44, 97-110.

Hutton, M. and Symon, C. J. (1986). The quantities of cadmium, lead, mercury and arsenic entering the U. K. environment from human activities. Sci. Total Environ. 57, 129-150.

Jackson, A. P. and Alloway, B. J. (1992). The transfer of cadmium from agricultural soils to the human food chain. In "Biogeochemistry of Trace Metals." (Ed. D. C. Adriano.) pp.109-158. (Lewis Publishers: London).

Jaakkola, A. (1977). Effect of fertilizers, lime and cadmium added to soil on the cadmium content of spring wheat. J. Sci. Agric. Soc. Finland 49, 406-414.

Jensen, A. and Bro-Rasmussen, F. (1992). Environmental cadmium in Europe. Rev. Environ. Contam. Toxicol. 125, 101-181.

John, M. K. (1972). Effect of lime on soil extraction and on availability of soil applied cadmium to radish and leaf lettuce plants. Sci. Tot. Environ. 1, 303-308.

John, M. K., VanLaerhoven, C. J. and Chuah, H. H. (1972). Factors affecting plant uptake and phyotoxicity of cadmium added to soils. Environ. Sci. Technol. 6, 1005-1009.

Kloke, A., Sauerbeck, D. R. and Vetter, H. (1984). The contamination of plants and soils with heavy metals and the transport of metals in terrestrial food chains. In "Changing Metal Cycles and Human Health". Ed. J. O. Nriagu) pp. 131-141. (Springer-Verlag: Berlin).

MacLean, A. J. (1976). Cadmium in different plant species and its availability in soils as influenced by organic matter and additions of lime, P, cadmium and zinc. Can. J. Soil Sci. 56, 129-138.

McLaughlin, M. J., Maier, N., Williams, C. M. J., Tiller, K. G. and Smart, M. K. (1993). Cadmium accumulation in potato tubers - occurrence and management. In "Proceedings 7th National Potato Research Workshop, Ulverstone, May 1993." (Ed. J. Fennell) pp.208-213. (Tasmanian Department of Primary Industry: Launceston).

McLaughlin, M. J., Tiller, K. G., Beech, T. A. and Smart, M. K. (1994a). Soil salinity causes elevated cadmium concentrations in field-grown potato tubers. J. Environ. Qual. 34, 1013-1018.

McLaughlin, M. J., Williams, C. M. J., McKay, A., Kirkham, R., Gunton, J., Jackson, K. J., Thompson, R., Dowling, B., Partington, D., Smart, M. K. and Tiller, K. G. (1994b). Effect of cultivar on uptake of cadmium by potato tubers. Aust. J. Agric. Res. 45, 1483-1495.

McLaughlin, M.J., Maier, N.A. Freeman, K., Tiller, K.G., Williams, C.M.J. and Smart, M.K. (1995a). Effect of potassic and phosphatic fertilizer type, phosphatic fertilizer cadmium content and additions of zinc on cadmium uptake by commercial potato crops. Fert. Res. (in press).

McLaughlin, M.J., K.G. Tiller, R. Naidu and D.P. Stevens.)1995b). The behaviour and environmental impact of contaminants in fertilizers. Aust. J. Soil Res. (in review).

Mench, M. and Martin, E. (1991). Mobilisation of cadmium and other metals from two soils by root exudates of Zea mays L., Nicotiana tabacum L. and Nicotiana rustica L. Plant and Soil 132, 187-196.

Merry, R.H. 1992. "Effects of farm management practices on cadmium uptake by pasture plants", Final Report (CS137) Meat Research Corporation, Canberra, Australia.

Merry, R. H. and Tiller, K. G. (1991). Distribution of cadmium and lead in an agricultural region near Adelaide, South Australia. Water, Air and Soil Poll. 57-58, 171-180.

Miller, J. E., Hassett, J. J. and Koeppe, D. E. (1976). Uptake of cadmium by soybeans as influenced by soil cation exchange capacity, pH, and available phosphorus. J. Environ. Qual. 5, 157-160.

Moen, J. E. T., Cornet, J. P. and Evers, C. W. A. (1988). Soil protection and remedial actions: criteria for decision making and standardisation of requirements. In "Contaminated Soil "88"". (Eds. J. W. Assink and W. J. van den Brink.) pp.1495-1503. (Kluwer: Dordrecht).

Mordvedt, J. J. (1987). Cadmium levels in soils and plants from some long-term soil fertility experiments in the United States of America. J. Environ. Qual. 16, 137-142.

Mordvedt, J. J., Mays, D. A. and Osborn, G. (1981). Uptake by wheat of cadmium and other heavy metal contaminants in phosphate fertilizers. J. Environ. Qual. 10, 193-197.

Mulla, D. J., Page, A. L. and Ganje, T. J. (1980). Cadmium accumulations and bioavailability in soils from long-term phosphorus fertilization. J. Environ. Qual. 9, 408-412.

NHMRC/ANZECC (1992). Australian and New Zealand Guidelines for the Assessment and Management of Contaminated Sites, January 1992. National Health and Medical Research Council and Australian and New Zealand Environment and Conservation Council, Canberra.

Nicholson, F.A., Jones, K.C., and Johnson, A.E. (1994). Effect of phosphate fertilizers and atmospheric deposition on long-term changes in the cadmium content of soils and crops. Environ.Sci. Technol. 28, 2170-2175.

OECD (1994). Organisation for Economic Cooperation and Development Risk Reduction Monograph No. 5. Cadmium. (Environment Directorate, Organisation for Economic Cooperation and Development, Paris).

Oliver, D. O., Schulz, J. E., Tiller, K. G. and Merry, R. H. (1993). The effect of crop rotations and tillage practices on cadmium concentration in wheat grain. Aust. J. Soil Res. 44, 1221-1234.

Oliver, D. O., Tiller, K. G., Conyers, M. K., Slattery, W. J., Merry, R. H. and Alston, A. M. (1994a). The effects of soil pH on cadmium concentration in wheat grain grown in south-eastern Australia. In "Third International Symposium on Plant-Soil Interactions at Low pH." (Ed. D. G. Edwards) (in press).

Oliver, D. O., Hannam, R., Tiller, K. G., Wilhelm, N. S., Merry, R. H. and Cozens, G. D. (1994b). The effects of zinc fertilization on cadmium concentration in wheat grain. J. Environ. Qual. (In press).

Oliver, D.P., Gartrell, G.W., Tiller, K.G., Correll, R., Cozens, G.D., and Youngberg, B.L. 1995. Differential responses of Australian varieties to cadmium concentration in wheat grain. Aust. J. Soil Res. (In press).

Page, A. L., Chang, A. C. and El-Amamy, M. (1987). Cadmium levels in soils and crops of the United States. In "Lead, Mercury, Cadmium and Arsenic in the Environment." (Eds. T. C. Hutchinson and Meema, K. M.) pp. 119-146. (John Wiley and Sons: New York).

Parker, D. R., Chaney, R. L. and Norvell, W. A. (1994). Chemical equilibrium models: Applications to plant nutrition research. In "Chemical Equilibrium and Reaction Models" (Eds. R. H. Loeppert et al.) Soil Science Society of America Spec. Publ.. (American Society of Agronomy: Madison). (in press).

Pepper, I. L., Bezdicek, D. F., Baker, A. S. and Sims, J. M. (1983). Silage corn uptake of sludge-applied zinc and cadmium as affected by soil pH. J. Environ. Qual. 12, 270-275.

Reuss, J. O., Dooley, H. L., Griffis, W. (1978). Uptake of cadmium from phosphate fertilizers by peas, radishes and lettuce. J. Environ. Qual. 7, 128-133.

Roberts, A.H.C., Longhurst, R.D., and Brown, M.W. 1994. Cadmium status of soils, plants and grazing animals in New Zealand. N.Z. Journal of Agricultural Research. 37, 119-129.

Rothbaum, H. P., Gogirel, R. L., Johnston, A. E. and Mattingley, G. E. G. (1986). Cadmium accumulation in soils from long-continued applications of superphosphate. J. Soil Sci. 37, 99-107.

Ryan, J.A., Pahren H.R. and Lucas J.B. (1982). Controlling cadmium in the human food chain: A rationale based on health effects. Env. Res. 28: 251-302.

Schroeder, H. A. and Balassa, J. J. (1963). Cadmium: uptake by vegetables from superphosphate and soil. Science 140, 819-820.

Sillanpaa, M. and Jansson, H. (1992). Status of cadmium, lead, cobalt and selenium in soils and plants of thirty countries. FAO Soils Bulletin 65, 194 pp. (Food and Agriculture Organisation of the United Nations: Rome).

Sparrow, L. A., Chapman, K. S. R., Parsley, D., Hardman, P.R. and Cullen, B. (1992). Response of potatoes (Solanum tuberosum cv. Russet Burbank) to band-placed and broadcast high cadmium fertiliser on heavily cropped krasnozems in north-western Tasmania. Aust. J. Expt. Agric. 32, 113-119.

Sparrow, L. A., Salardini, A. A. and Bishop, A. C. (1993a). Field studies of cadmium in potatoes (Solanum tuberosum L.). I. Effects of lime and phosphorus on cv. Russet Burbank. Aust. J. Agric. Res. 44, 845-853.

Sparrow, L. A., Salardini, A. A. and Bishop, A. C. (1993b). Field studies of cadmium in potatoes (Solanum tuberosum L.). II. Response of Cvv. Russet Burbank and Kennebec to two double superphosphates of different cadmium concentrations. Aust. J. Agric. Res. 44, 855-861.

Sparrow, L.A., Salardini, A.A., and Johnsone,J. 1994. Field studies of cadmium in potatoes(Solanum tuberosum L.) III.Response of cv.Russet Burbank to sources of banded potassium.Aust.J.Agric.Res. 45, 243-249.

Stenhouse, F. (1991). The 1990 Australian Market Basket Survey Report. National Health and Medical Research Council/National Food Authority. (Australian Government Printing Service: Canberra).

Stenhouse, F. (1992). The 1992 Australian Market Basket Survey. National Food Authority. (Australian Government Printing Service: Canberra).

Thomas, B., Roughan, J. A. and Watters, E. D. (1972). Lead and cadmium content of some vegetable foodstuffs. J. Sci. Fd. Agric. 23, 1493-1498.

Tiller, K. G. (1991). Determining background levels. In "The Health Risk Assessment and Management of Contaminated Sites" (Eds. O.E. Saadi and A. Langley) pp 98-101. (S.A. Health Commission: Adelaide).

Tiller, K. G., Oliver, D. P., McLaughlin, M. J., Merry, R. H. and Naidu, R. (1994). Managing cadmium contamination of agricultural land. Adv Environ. Sci. (in press).

Tjell, J. C., Hansen, J. A., Christensen, T. H. and Hovmand, M. F. (1981). Prediction of cadmium concentrations in Danish soils. In "The Second European Symposium on Characterization, Treatment and Use of Sewage Sludge, Vienna 20-24 October, 1980." (Eds. P. L"Hermite and H. Ott) pp 650-664. (D. Reidel: London).

Weigert, P., Muller, J., Klein, H., Zufelde, K.P. and Hillebrand, J. (1984). Arsen, Blei, Cadmium und Quecksilber in und auf Lebensmitteln. ZEBS Hefte 1.(Federal Republic of Germany).

White, M. C., and Chaney, R. L. (1980). Zinc, cadmium and Mn uptake by soybean from two zinc- and cadmium-amended coastal plain soils. Soil Sci. Soc. Am. J. 44, 308-313.

Wiersma, D., van Goor, B. J., van der Veen, N. G. (1986). Cadmium, lead, mercury and arsenic concentrations in crops and corresponding soils in the Netherlands. J. Agric. Food Chem. 34, 1067-1074.

Willaert, G. and Verloo, M. (1992). Effects of various nitrogen fertilizers on the chemical and biological activity of major and trace elements in a cadmium contaminated soil. Pedologie 43, 83-91.

Williams, C. H. and David, D. J. (1973). The effect of superphosphate on the cadmium content of soils and plants. Aust. J. Soil Res. 11, 43-56.

Williams, C. H. and David, D. J. (1976). The accumulation in soil of cadmium residues from phosphate fertilizers and their effect on the cadmium content of plants. Soil Sci. 121, 86-93.

Williams, C. H. and David, D. J. (1977). Some effects of the distribution of cadmium and phosphate in the root zone on the cadmium content of plants. Aust. J. Soil Res. 15, 59-68.

Wolnik, K. A., Fricke, F. L., Capar, S. G., Braude, G.L., Meyer, M. W., Satzger, R. D. and Bonnin, E. (1983). Elements in major raw agricultural crops in the United States. 1. Cadmium and lead in lettuce, peanuts, potatoes, soybeans, sweet corn and wheat. J. Agric. Food Chem 31, 1240-1244.

Wolnik, K. A., Fricke, F. L., Capar, S. G., Meyer, M. W., Satzger, R. D., Bonnin, E. and Gaston, C. M. (1985). Elements in major raw agricultural crops in the United States. 3. Cadmium, lead and eleven other elements in carrots, field corn, onions, rice, spinach and tomatoes. J. Agric. Food Chem. 33, 807-811.

Evidence for the Leaching of Surface Deposited Cadmium in Agricultural Soils

Fiona A. Nicholson,[1] Kevin C. Jones[2] and A. E. "Johnny" Johnston[3]

[1]ADAS Gleadthorpe, Meden Vale, Mansfield, Nottinghamshire,
[2]IEBS, Lancaster University, Lancaster
[3]IACR, Rothamsted, Harpenden, Hertfordshire
United Kingdom

Abstract

Cadmium enrichment between 1913 and 1991 of a soil profile (0-22.5 cm) under permanent grassland indicates that there has been appreciable leaching of cadmium deposited on the surface both in atmospheric deposition and superphosphate fertiliser. Cadmium in each soil layer (2.5 cm) was determined on the limed (pH 6.4) and unlimed (pH 4.8) soils, with and without annual applications of superphosphate, in the Park Grass experiment at Rothamsted Experimental Station. Construction of a cadmium budget for these soils suggested that leaching was an important pathway of loss of cadmium added in superphosphate and was not affected by the difference in soil pH (1.6 units). Cadmium in soils was determined by GFASS following a nitric acid digestion.

Short term batch and column experiments were used to determine the effect of the soltion pH of simulated rainfall on the extraction and leaching of cadmium from soil. The pH of the input solution determined the amount of cadmium released in each sequential extraction, but the total amount of cadmium extracted depended on the amount of acidity added rather than the pH of the extracting solution. Some initial predictions of cadmium soil residence times with acid deposition of different pH suggest that an increase in acidity of 1 pH unit would lead to a tenfold decrease in the residence time of cadmium in these soils.

Introduction

Cadmium enters agricultural soils mainly through atmospheric deposition and the application of phosphatic fertilisers and sewage sludges, and is lost through leaching, crop offtake and soil erosion. Inputs appear to be greater than losses and, in the long term, cadmium is accumulating to levels significantly above natural concentrations (Rothbaum et al., 1986; Jones et al., 1987). It is therefore important to understand the mechanisms governing the removal of cadmium from soils and how these may be influenced by changing environmental conditions.

The literature presents conflicting evidence about the importance of cadmium losses through leaching. McGrath and Lane (1989) accounted for about 80 per cent of the cadmium added to soil in sewage sludge over a 20 year period. The top 25 cm of soil, which was high in organic matter, had pH 6.5 and there was little evidence to suggest cadmium enrichment in the deeper soil layers. McGrath (1987) measured a low concentration of cadmium in the soil water and suggested that leaching was unlikely to be a significant route of loss. Davis et al. (1988) found that some cadmium added in sludge to grassland soils had moved to a depth of 10 cm 3-5 years after application, although 80 per cent of the cadmium remained in the top 5 cm. When cadmium was added to arable soils in phosphate fertilisers over a period of 20 years, Williams and David (1976) showed that about 80 per cent was retained in the cultivated horizon except in the case of a light, sandy soil from which 50 per cent of the cadmium was lost. Rothbaum et al. (1986) found only about 50 per cent of the cadmium added to a New Zealand grassland soil (pH 5.9) had been retained in the top 22.5 cm. Because crop offtake (ryegrass) is thought to account for <10 per cent of the total cadmium added in phosphorous-fertilisers (He and Singh, 1994), these data imply that some added cadmium may be lost through leaching. Few estimates of such losses have been attempted, but Bowen (1975) suggested an annual loss of 5 g cadmium/ha as an average for UK soils and Tjell and Christensen (1992), using a mass-balance approach for Danish agricultural soils, suggested 2 g cadmium/ha/yr from sandy soils and 1 g cadmium/ha/yr from clay soils. Leaching of cadmium will depend on the solubility of cadmium compounds and complexes added to soil and the ability of soil constituents to retain cadmium in these various forms. It will also depend on the residence time of drainage water in soil. Thus cadmium leached from surface soils may be retained in deeper horizons or be removed from the soil profile.

At the same time as they are accumulating cadmium, soils may be acidifying. Acidification arises from natural processes within the soils or from acidifying inputs from atmospheric deposition or nitrogen fertilisers. It is generally accepted that increasing the acidity of a soil will increase the solubility of cadmium in the soil solution. As a result, the mobility of cadmium in the soil profile also increases, although this depends not only on pH but also on other soil properties, such as cation exchange capacity and factors like the amount of soil organic matter.

The effect of acid deposition on the leaching of cadmium from acidic, forest soils has been investigated in some detail. For example, Tyler (1978) applied artificial rainwater of different pH to polluted and control forest soils and calculated the number of years for a 10 per cent decrease in total soil cadmium in the mor horizon through leaching. Rasmussen et al. (1988), using soil cores from an acidic spruce forest soil, applied rainfall at pH 2.8-3.3 over a period of 9 months. Short term increases in acidity of precipitation increased the mobility of cadmium in these soils. Similarly, Berthelsen et al. (1994) applied artificial acid rain to pine forest plots over a period of 7-8 years, and concluded that substantial amounts of cadmium were leached from the O horizon, probably into groundwater.

Questions remain to be answered, however, in relation to the effect of acid rain inputs on leaching of cadmium from agricultural soils of different texture and organic matter content. Such soils may have received cadmium either from atmospheric deposition or the addition of phosphate fertilisers or sludges. Agricultural soils may also be subject to a change in use. Long-term set-aside or afforestation almost invariably lead to a cessation of liming and a fall in soil pH (Johnston et al., 1986; Stigliani and Salomons, 1993).

The importance of leaching as a route for cadmium loss over a long time period (78 years) was shown using soils from the Park Grass experiment at Rothamsted Experimental Station. Laboratory experiments determined the effect of acid deposition on cadmium extraction and leaching with two other agricultural soils. Both sets of data were used to develop a model to predict the release of cadmium from agricultural soils based on assumptions about the acidity of rainfall and the level of cadmium contamination in agricultural soils.

Materials and methods

The Park Grass experiment was started in 1856 on an area that had already been under permament grassland for at least 200 years (Warren and Johnston, 1964). Some data on the cadmium content of soil from plot 4/1 which has received superphosphate annually since 1859, and from plot 3 which has never received fertilisers are shown in Table 1. In 1903, the plots were halved; one half received applications of lime every fourth year and by 1991 had a soil pH of 6.5 (phosphorous-treated) and 6.4 (control); the other halves have continued without lime and in 1991 had reached a pH of 4.9 (phosphorous-treated) and 4.8 (control). In 1913, soil cores were taken from the unlimed sections of the control and superphosphate-treated Park Grass plots. The cores were divided into 2.5 cm layers, air dried, ground to pass a 2 mm sieve and stored in glass jars. Soil cores were taken in 1991 from both the limed and unlimed sections of the plots and prepared in the same way. The soils were recently subsampled and the cadmium content determined using Zeeman correction GFAAS following a nitric acid digestion on two replicate samples. Although nitric acid digestion does not solubilise silicates, it was assumed that applied cadmium will not be held within the crystal lattice and therefore that this method gives a good estimate of the amount of cadmium involved in sorption or leaching processes (Boekhold and Van der Zee, 1992).

The effects of the acidity of input solutions simulating rainfall were investigated by both a sequential batch extraction and the leaching of repacked soil columns using two soils. The soils, from commercial farms, were a silt/loam grassland soil (pH 5.6; total cadmium = 0.38 mg/kg) and a sandy loam arable soil (pH 6.6; total cadmium = 1.2 mg/kg). Prior to use, the soils were air dried, ground and sieved to <2 mm. Artifical acid rain was made up from a mixture of 0.1M H_2SO_4 and 0.1M HNO_3 (Primar grade, made up in Milli-Q water) in a ratio of 7:3 by volume (Harter, 1989), to give solutions of pH 1.0-4.0.

In the batch experiment, 80 ml of simulated acid rain of the appropriate pH was added to 3 replicate 10 g soil samples in acid washed polythene bottles before being placed on a rotary shaker for 24 hours. The pH of the solutions was then recorded before being centrifuged and the supernatant decanted and analysed for cadmium by GFAAS. Additional 80 ml aliquots of acid rain were added to the soil, and the process was repeated 6 times.

In the column experiment, 100 g of soil was placed in a 35 mm diameter leaching column and acid rain of the appropriate pH was added to bring the soil to field capacity. The columns were left to equilibrate for 24 hours. Acid rain was then applied slowly to the soil and the leachate collected in 100 ml aliquots in acid washed polythene bottles. The pH of each aliquot of leachate was measured prior to cadmium analysis as above.

Results and discussion

Leaching losses of cadmium from Park Grass soils

Figures 1 and 2 show the soil cadmium concentrations in 1913 and 1991, at different depths, of the unlimed, control and the phosphorous-treated plots respectively. Detailed data for the cores taken from the limed sections of the plots in 1991 are not shown. In these soils, which have been undisturbed for many years, the distribution of cadmium can only be affected by downward movement in the soil solution, translocation by plant roots or by earthworm activity. Visual evidence suggests little earthworm activity and root translocation is unlikely to be significant at depths greater than a few centimetres. Thus changes in vertical distribution can be mostly accounted for by downward movement of cadmium in drainage water.

Average soil cadmium concentrations were calculated using appropriate soil weights for each horizon and are compared with those obtained by other researchers in Table 1. The increase in soil cadmium (0-22.5 cm) on the control plots between 1913 (0.21 mg cadmium/kg) and 1991 (0.29 mg cadmium/kg) was due to the accumulation of cadmium from atmospheric deposition. Figure 1 shows that cadmium from this source is not retained solely in the surface layers of the soil, but moves down and enriches the soil profile to a depth of about 25 cm. The movement may have been aided by increased inputs of Cl^- in deposition which formed labile cadmium complexes in the soil or by a decrease in rainfall pH since 1913. Similar movement has been observed over a 12 year period in a highly polluted woodland soil (Martin and Bullock, 1994).

On the phosphorous-treated plots (Figure 2) the increase in soil cadmium between 1913 (0.24 mg cadmium/kg) and 1991 (0.36 mg cadmium/kg) was due both to atmospheric deposition and to the use of commercial superphosphates containing cadmium. Again, the same movement of cadmium down the soil profile to a depth of about 25 cm is observed on these plots.

Table 1

Reported soil cadmium concentrations (mg/kg) from the Park Grass experiment

Date	Control	phosphorous-treated	Source
1876	0.20	0.17	Rothbaum et al. (1986)
1876	0.19		Jones et al. (1987)
1913	0.21	0.24	This study
1959	0.18	0.31	Rothbaum et al. (1986)
1959	0.21		Jones et al. (1987)
1975	0.33	0.46	Rothbaum et al. (1986)
1976	0.31	0.44	Rothbaum et al. (1986)
1982	0.16	0.29	Rothbaum et al. (1986)
1984	0.27	0.41	Jones et al. (1987)
1991	0.29	0.36	This study

Figure 1

Soil cadmium concentrations for the unlimed Park Grass control plot in 1913 and 1991

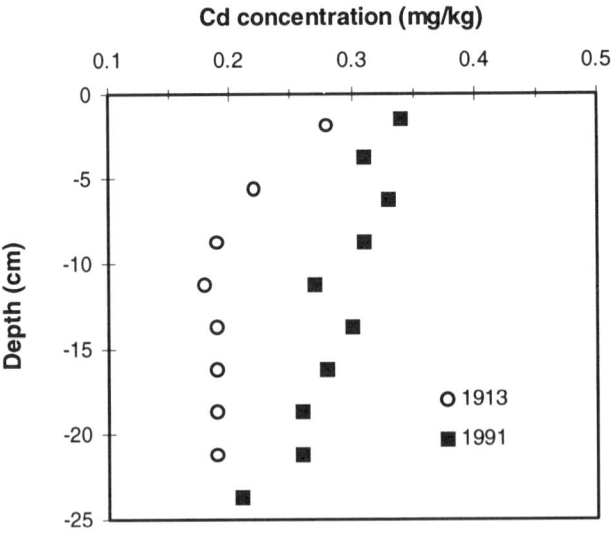

Figure 2

Soil cadmium concentrations for the unlimed unlimed Park Grass phosphorous-treated plot in 1913 and 1991

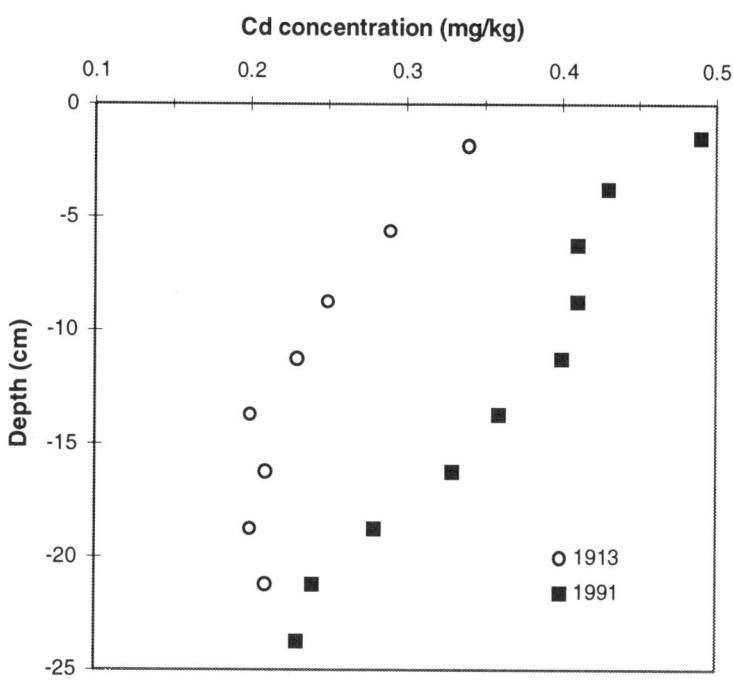

An attempt to estimate the amount of cadmium lost through leaching was made by constructing a cadmium budget for the soils (Table 2). Because there were some differences in the soil cadmium concentrations obtained in recent years for these soils (Table 1), it was decided to use an average value for the 16 year period 1975-1991 to establish present day soil cadmium concentrations, and hence to base the calculations on the 70 years between 1913 and 1983. An average soil weight of 2676 t/ha was used to calculate the total amount of cadmium in the present day 0-22.5 cm profiles. The 1913 total was then subtracted from the present day total to give the amount of cadmium which had accumulated in the soil of each plot over this time period. Inputs from fertilisers were calculated using data from Nicholson et al. (1994) for the cadmium content of phosphates applied between 1925 and 1988 (mean cadmium content = 10 mg/kg) and averaged 4.9 g cadmium/ha/yr. Cadmium removed by crop offtake depended on crop yield which varied between the plots, between harvests and over time (Nicholson et al., 1994). Losses from soil erosion or lateral movement are unimportant on these plots which have remained undisturbed for at least 300 years and are on level ground.

The amount of cadmium input from atmospheric deposition was more difficult to derive as no direct measurements of deposition rates have been made for these plots. The average increase in soil cadmium observed on Rothamsted control plots (Jones et al., 1987) was 3.2 g cadmium/ha/yr which is similar to the rate of 3 g cadmium/ha/yr assumed in a study by Hutton and Symon (1987) and considered a representative value for cadmium deposition to agricultural land in the European Union (Alloway, 1990). However, this measured increase results from a net input of cadmium ie. it is the increase observed after any losses through leaching or plant uptake. It is therefore likely that the average atmospheric input between 1913 and 1991 was higher than 3.2 g cadmium/ha/yr, and this value was used as a minimum when constructing the budget; 5.4 g cadmium/ha/yr was used as a maximum value for atmospheric inputs to the plots (Jones et al., 1987). By assuming that aerial deposition was the same to both plots, the effect of the phosphorous-treatment could be estimated.

Table 2

A cadmium budget for the 0-22.5 cm horizon of Park Grass soils (1913 -1983)

	Untreated (g/ha)	P-treated (g/ha)	Difference (g/ha)
Accumulated in soil	179	409	230
Estimated inputs			
Atmospheric deposition	250-421	250-421	0
Phosphate fertilisers	0	386	386
Estimated losses			
Crop offtake	24	56	32
Leaching from top 22.5 cm	47-218	171-342	130
Estimated leaching rate (g/ha/yr)	0.7-3.1	2.4-4.9	1.8

Table 2 shows that of the 386 g cadmium/ha added in phosphorous-fertilisers, 230 g cadmium/ha (60 per cent) has been retained in the soil. This is slightly higher that the 50 per cent retention estimated by Rothbaum et al. (1986) for a New Zealand grassland soil. It is not possible to establish whether the cadmium lost by leaching from the top 22.5 cm has been adsorbed by the deeper subsoil as some authors have suggested (Miller et al., 1983), although this seems unlikely as Rothbaum et al. (1986) found little difference in the cadmium content of subsoils (23-40 cm) at Rothamsted after 100 years. It appears, therefore, that cadmium, once removed from the topsoil, has remained in solution in the drainage water. In either case, the data seem to indicate that cadmium has been lost at a higher rate from the phosphate treated plot compared with the untreated plot. This imples that cadmium added in phosphatic fertilisers may be retained less strongly by the topsoil than that from atmospheric deposition and may be more likely to be affected by changes in soil acidity or acid deposition.

Cadmium concentrations at all depths in the 1991 soil cores from the limed plots were little different to those from the unlimed plots. Average soil cadmium concentrations were 0.28†mg cadmium/kg (untreated, limed), 0.29 mg cadmium/kg (untreated, unlimed), 0.37†mg cadmium/kg (phosphorous-treated, limed), 0.36 mg cadmium/kg (phosphorous-treated, unlimed). Thus, on both the untreated and the phosphorous-treated soils, very little extra cadmium has been leached from the limed and unlimed half of each treatment. It appears that for this grassland soil, which is high in organic matter, the mobility of cadmium added either from atmospheric deposition or through the use of phosphate fertilisers has not been differentially increased by the increased acidity on the plot without applications of lime and now with a pH of 4.8.

Effect of precipitation pH on cadmium extraction

Figure 3 shows the pH of the solution after successive extractions of the grassland soil with simulated acid rain of different pH. The pH of the supernatant solution eventually approached that of the input solution. Similar results were obtained for the arable soil and for the column experiments (not shown). In both cases the soil had some capacity to neutralise added acidity.

For input solutions of higher pH (3.0-4.0), the pH of the solution after extraction was well above the initial pH showing that the soil was able to buffer this added acidity to a large extent. The cadmium concentration in the extracted solution remained almost constant (< 5 ng cadmium/g soil), which may explain why the fall in soil pH on the unlimed Park Grass soils had little effect on the amount of cadmium retained. With more acid extractants (pH < 3.0) the pH of the supernatant decreased rapidly, approaching the initial pH of the input solution, and much more cadmium was removed from the soil. The rate of release of cadmium from the soil was a function of the acidity introduced with the leaching solution; more cadmium was released the more acid the solution applied to the soil.

The relationship between the cumulative cadmium extracted and the total acidity added for the grassland soil in the batch and column experiments (Figures 4 and 5), shows that much cadmium was extracted over a narrow range of total added acidity. For the batch experiment, most cadmium was extracted between 100-500 µeq H^+/g soil for the grassland soil (Figure 4) and between 100-800 µeq H^+/g soil for the arable soil (not shown). In the column experiment most cadmium was extracted between 50 and 200 µeq H^+/g soil for the grassland soil (Figure 5) and 200-500 µeq H^+/g soil for the arable soil (not shown). The total cadmium extracted in the batch experiments (about 0.25 mg cadmium/kg or 66 per cent of the total grassland soil cadmium and 1.1 mg cadmium/kg or 92 per cent of the total arable soil cadmium) was greater than in the column experiment (about 0.19 mg cadmium/kg or 50 per cent of the total grassland soil cadmium and 0.93 mg cadmium/kg or 78 per cent for the arable soil) probably because repeated extraction with aliquots of simulated acid rain was very efficient at removing cadmium.

The pattern of extraction was dependent on the total quantity of H^+ added to the soil and it was speculated that the same shaped curve would be produced if rain of a higher pH in larger volumes was to be added for a longer time. Based on this assumption and a knowledge of the average amount and pH of rainfall, it is possible to make some initial predictions of how long it would take to release 50 per cent of the acid leachable cadmium from agricultural soils.

If through drainage is assumed to be 250 mm per year, this equates to a volume of 250 l/m² annual percolation. If the rain has pH 4.5, then 250 l delivers 7,900 µeq H⁺ to 1 m² soil. Assuming the top 23 cm of soil weighs 230 kg (based on a density of 1 g/cm³) then 7,900 µeq H⁺ are applied to 230 kg soil or 0.034 µeq H⁺/g/year. From Figure 5, 50 per cent of the extractable cadmium would be released after about 130 µeq H+/g soil have been added, which would take about 3800 years. For the arable soil (data not shown), 300 µeq H⁺/g soil would be required to release 50 per cent of the cadmium which would take about 8800 years. If the acidity of the rain was to increase, the release would occur sooner as shown in Table 3.

Some previous estimates have been made of cadmium residence times in soils. Bowen (1979) quotes a range of 75-38 years, whilst Imura et al. (1977) suggest 5-1,100 years and Nriagu (1980) gives a figure of 3,000 years. McGrath (1987) assumed that leaching losses from the sludged soils he studied were low and estimated that it would take 7,500 years to remove 100 per cent of the cadmium added in sewage sludge by crop offtake alone. A study by Tyler (1978) on the organic mor horizon of spruce forest soils gave shorter residence times than this study using a similar range of precipiation pHs.

Figure 3

Change in pH of the batch extracted solution with amount of added acidity for a silt/loam grassland soil

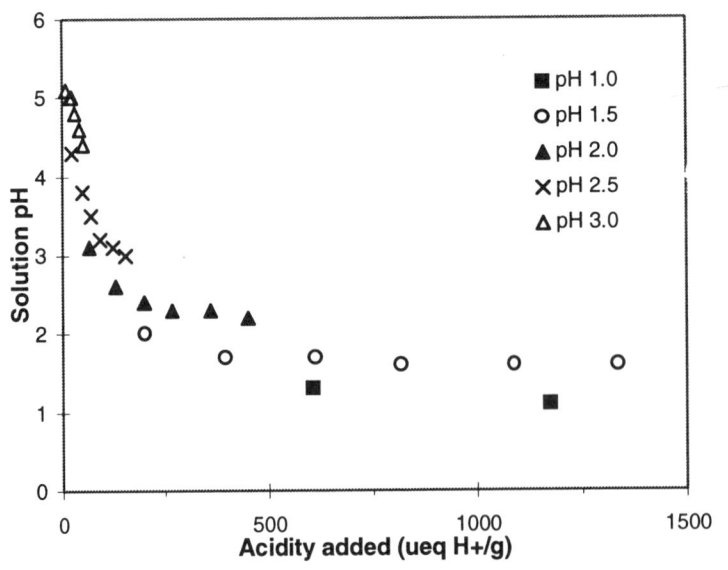

Figure 4

Cadmium extracted from silt/loam grassland acid rain (batch experiment)

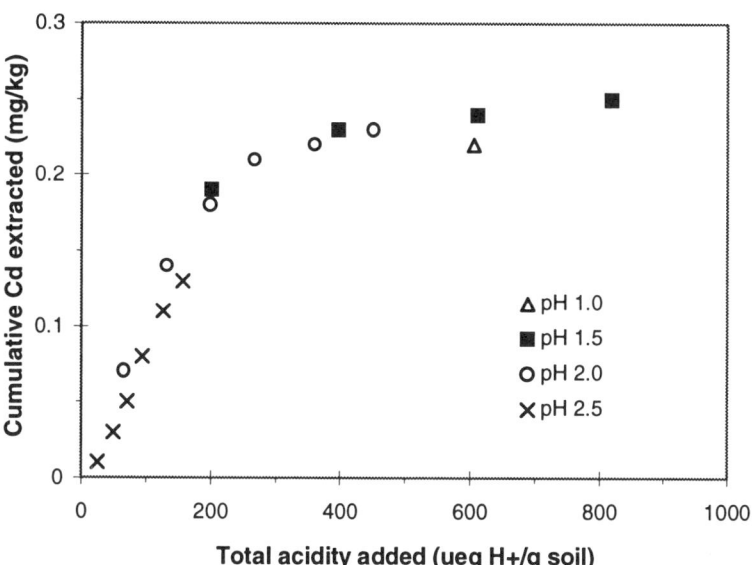

Figure 5

Cadmium extracted from silt/loam grassland soil by acid rain (column experiment)

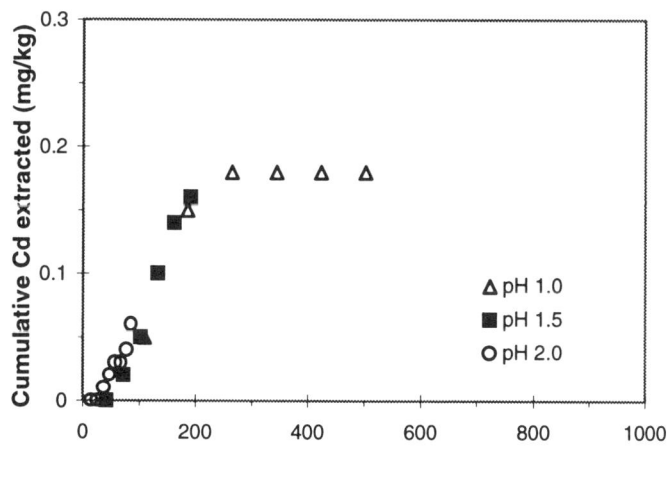

It is not clear whether the data obtained from this study are directly comparable with other literature values because the simple model used has not taken into account future inputs of cadmium to the soil nor has it considered losses through crop uptake, soil erosion, lateral movement or downward movement with particulate material. There may also have been substantial differences in soil texture or buffering capacity as well as different annual percolation rates. Nevertheless, they illustrate one of the potential consequences for ordinary agricultural soils of failing to control emissions which lead to acid rain formation and hence soil acidification and mobilisation of metals.

There are many problems with an attempt such as this to transfer experimentally assessed leaching rates to residence times in the field. The formation of soluble cadmium-organic complexes may accelerate the leaching rate as decomposition of organic matter occurs (Wang and Stumm, 1987). Soil properties such as cation exchange capacity and the amount and type of base cations present will undoubtedly exert a significant influence on the ability of the soil to retain cadmium following the addition of H^+ in acid deposition, and factors such as the composition of the input solution and the quantity of rainfall will also play an important part.

Table 3

Number of years to the release of 50 per cent of the acid extractable cadmium in two soils

Precipitation pH	Grassland soil (years)	Arable soil (years)
4.5	3800	8800
4.0	1200	2800
3.5	380	880
3.0	120	280
2.5	38	88

Conclusions

The first part of this study suggests that an organic rich, grassland soil, retains 60 per cent of the cadmium added in phosphorous-fertilisers and that leaching may be an important pathway for the loss of cadmium entering the soil from this source. Liming, whilst controlling soil acidification, did not affect the downward movement of cadmium or its retention in this soil. Thus the natural processes of soil acidification which have decreased the soil pH by 1.6 pH units over 78 years, have not resulted in an increase in the mobility of cadmium either from atmospheric deposition or from phosphorous-fertilisers.

In the second part of this study, the effect of acid deposition on cadmium mobilisation was examined in a laboratory study. For the two soils used, the pH of the input solution determined the rate of release of cadmium from the soil. However, the pattern of cadmium extraction depended on the total amount of acidity added to the soil. Some initial predictions were made of the 50 per cent residence time of cadmium in the soils at different deposition pH. The results suggest that an increase in rain acidity of 1 pH unit would lead to a tenfold decrease in the 50 per cent residence time of cadmium in the soils, although it is recognised that many problems arise when transferring these predictions to the field.

Acknowledgements

The authors would like to thank the UK Biotechnology and Biological Sciences Research Council (BBSRC) for funding this work.

References

Alloway, B. J. 1990. Heavy Metals in Soils. Blackie, Glasgow and London, 105p.

Berthelsen, B. O., Ardal, L., Steinnes, E., Abrahamsen, G., Stuanes. A. O. 1994. Mobility of heavy metals in pine forest soils as influenced by experimental acidification. *Water, Air, Soil Pollut.*, 73, 29-48.

Boekhold, A. E., Van Der Zee, S. E. A. T. M. 1992. Significance of soil chemical heterogeneity for spatial behaviour of cadmium in field soils. *Soil Sci. Soc. Am. J.*, 56, 747-754.

Bowen, H. J. M. 1975. Soil pollution. *Educ. Chem.* 72-76.

Bowen, H. J. M. 1979. *Environmental Chemistry of the Elements*. Academic Press, London.

Davis, R. D., Carlton-Smith, C. H., Stark, J. H., Campbell, J. A. 1988. Distribution of metals in grassland soils following surface applications of sewage sludge. *Environ. Pollut.*, 49, 99-115.

Harter, P. 1989. *Acid Deposition-Ecological Effects on Soils and Forests*, IEA Coal Research, London.

He, Q. B., Singh, B. R. 1994. Crop uptake of cadmium from phosphorus fertilisers: I. Yield and cadmium content. *Water, Air, Soil Pollut.*, 74, 251-265.

Hutton, M., Symon, C. 1987. Sources of cadmium discharge to the UK environment. In P. J. Coughtrey, M.†H. Unsworth (eds): *Pollutant Transport and Fate in Ecosystems*. Blackwell Scientific Publications, Oxford, p. 223.

Iimura, K., Ito, H., Chino, M., Marishita, T., Hiruta, H. 1977. *Proceedings of the International Seminar SEFMIA*, Tokyo. pp 35 -366.

Johnston, A. E., Goulding, K. W. T. and Poulton, P. R. 1986. Soil acidification during more than 100 years under permanent grassland and woodland at Rothamsted. *Soil Use and Management*, 2, 3-10.

Jones, K. C., Symon, C. J., Johnston, A. E. 1987. Retrospective analysis of an archived soil collection. II. *Cadmium. Sci. Tot. Environ.*, 67, 75-89.

Martin, M. H., Bullock, R. J. 1994. The impact and fate of heavy metals in a contaminated woodland ecosystem. In P. J. Coughtrey, M. H. Unsworth (eds): *Pollutant Transport and Fate in Ecosystems*. Blackwell Scientific Publications, Oxford, pp. 327-365.

McGrath, S. P. 1987. Long-term studies of metal transfers following application of sewage sludge. In P. J. Coughtrey, M. H. Unsworth (eds): *Pollutant Transport and Fate in Ecosystems*. Blackwell Scientific Publications, Oxford, pp.301-307.

McGrath, S. P., Lane, P. W. 1989. An explanation for the apparent losses of metals in a long-term field experiment with sewage sludge. *Environ. Pollut.*, 60, 235- 256.

Miller, W. P., McFee, W.W., Kelly, J. M. 1983. Mobility and retention of heavy metals in sandy soils. *J. Environ. Qual.*, 12, 579-584.

Nicholson, F. A., Jones, K. C., Johnston, A. E. 1994. Effect of phosphate fertilisers and atmospheric deposition on long-term changes in the cadmium content of soils and crops. *Environ. Sci. Technol.*, 28, 2170-2175.

Nriagu, J. O. 1980. *Cadmium in the Environment. Part 1: Ecological Cycling*, Wiley-Interscience, New York. pp 2-12

Rasmussen, L., Von Freiesleben, N. E., Jorgensen, P. 1988. Leaching of ions from a forested typic udipsamment by acidified throughfall, Denmark. *Geoderma*, 43, 33-47.

Rothbaum, H. P., Goguel, R. L., Johnston, A. E., Mattingly, G. E. G. 1986. Cadmium accumulation in soils from long-continued applications of superphosphate. *J. Soil Sci.*, 37, 99-107.

Stigliani, W. M., Salomons, W. 1993. Our father's toxic sins. *New Sci.* Vol. 140, No.1903, 38-42.

Tjell, J. C., Christensen, T. H. 1992. Sustainable management of cadmium in Danish agriculture. In J-P Vernet (ed.): *Impact of Heavy Metals on the Environment*. Elsevier, Amsterdam, 273-286.

Tyler, G. 1978. Leaching rates of heavy metal ions in forest soil. *Water, Air, Soil Pollut.*, 9, 137-148.

Wang, Z., Stumm, W. 1987. Heavy metal complexation by surfaces and humic acids: a brief discourse on assessment by acidimetric titration. *Neth. J. Agric*. Res., 35, 231-240.

Warren, R. G., Johnston, A. E. 1964. Report of the Rothamsted Experimental Station for 1963.

Williams, C. H., David, D. J. 1976. The accumulations in soil of cadmium residues from phosphate fertilisers and their effects on the cadmium content of plants. J. *Soil Sci.*, 121, 86-93.

REPORT OF SESSION D

UPTAKE INTO CROPS AND BIOAVAILABILITY

The papers in this Section were presented during Session D.

Initial considerations

The group decided not to discuss sewage sludges and localised pollution, e.g. from smelters, in detail. Therefore this discussion relates to fertilizer cadmium, but some of the comments will apply to manure cadmium.

The group defined management option time scales as: short term (one to five years) and long term (>50 years).

Phytoavailability was defined as the availability of cadmium in soil to plants.

Bioavailability was defined as the availability of cadmium in foods to mammals.

Conclusions

The group recognised the need to minimise the accumulation of cadmium in soils because this may cause an increase in cadmium concentrations in crops. They also thought it necessary to take a dispassionate view of cadmium concentrations in crops. Crop cadmium concentrations can be decreased by identifying and applying appropriate regional packages of good management practices based on a prescription approach as defined below. Such an approach must recognise the need to maintain the economic viability of farming or introduce appropriate compensation.

If levels of soil cadmium continue to increase with an increased risk of cadmium transfer to the food chain, alternative strategies to those outlined below must be sought. For example, one approach would be to assess whether crops can be genetically adapted to decrease cadmium uptake. Success in minimising cadmium uptake by crops through genetic engineering should not be an excuse for ignoring strategies to diminish cadmium input to soils.

Management factors

Cadmium addition to soil

Impact on crop cadmium depends upon cadmium source, cadmium/zinc ratio and soil characteristics (see below) rather than the amount of cadmium alone. Experimental evidence suggests that the short term impact of cadmium additions on plant cadmium concentrations is insignificant. Others thought there might be significant effects.

Cadmium inputs in fertilizer, manure or other sources could be managed by reducing the amount in the source material or reducing the amount of fertilizer or manure phosphate added to soil. There is a danger of decreasing phosphate applications decreasing yields.

Cadmium concentration in soil

In any given soil, cadmium in soil solution is related to accumulation of cadmium from fertilizer, atmospheric deposition and background levels of total cadmium in soil. Soil solution and crop cadmium concentrations are related, all other factors being equal, for any given soil. This relationship is less clear for other sources. Across soils and regions, crop and soil cadmium concentrations are less well related.

Cadmium concentrations in soil cannot be readily decreased. Thus crop cadmium concentrations may need to be managed by regulation and/or enforcement of where different crop species are grown. The use of hyperaccumulator crops for removal of cadmium is not an economic option at present.

pH

In general, in non-alkaline soils lowering pH leads to higher cadmium uptake by crops. Soil acidity has both a short and long term impact on crop cadmium concentrations. A lack of response of crop cadmium concentrations to applications of liming materials has been noted in some situations.

Soil pH can be maintained or increased by the use of liming materials, and this is economic in some systems. In many areas, particularly less developed countries, addition of liming materials is not economically feasible and agriculture depends on aluminium-tolerant crops. The development of aluminium-tolerant cultivars may lead to higher crop cadmium concentrations unless steps are taken to minimise cadmium uptake characteristics.

Nitrogen

In the short term, the amount, form and placement of applied nitrogen may affect cadmium uptake, possibly through impacts on root growth or the chemistry of cadmium in soil or rhizosphere. Nitrogen applications can be managed without major cost, but all crops need nitrogen for economic yield. The costs involved in changing nitrogen management are minimal.

Phosphates

In some species there is evidence that phosphate application may cause increases in crop cadmium uptake in the short term. In the long term, when soil phosphate levels have been increased to sufficiency levels, further phosphate additions will have little effect on cadmium availability. There are no major costs involved in changing phosphate management strategies, but all crops need an appropriate amount of available soil phosphate to achieve economic yields.

Zinc

In considering applications of zinc to reduce crop cadmium concentrations, there is a need to distinguish between zinc-deficient and zinc-sufficient soils. There is conclusive evidence that alleviation of zinc deficiency reduces crop cadmium concentrations in the short term. In the long term, repeated zinc applications would be required to avoid deficiency. This can be achieved at low cost.

There is some evidence that addition of zinc to zinc-sufficient soils reduces crop cadmium concentrations in the short term and this may be economic depending on rate of zinc used and the value of the crop. There is no information on long term effects.

Selection of crop

There are large differences in cadmium uptake between species. Crop type can therefore be changed, or the use to which a crop is put may be changed (e.g. from food to animal feed). Any restriction on the crop that a farmer can grow may have a major impact on the financial viability of individual farms.

Cultivar

There are differences in cadmium uptake between cultivars, and new cultivars have been identified which have low cadmium accumulation characteristics. Some countries are more advanced than others in terms of using this management option. Success thus far has been achieved with sunflower, durum wheat and potatoes. However, other favourable crop attributes (agronomic performance, disease resistance, nutritional properties, etc.) must not be sacrificed whilst seeking to breed for low cadmium uptake.

Crop rotation

In the long term, cadmium uptake may be changed by changing crop rotations. This effect may result through changes in other soil factors such as soil pH, organic matter, soil pathogens, etc. These changes, however, may decrease profitability if substituted crops do not have similar value.

Salinity

Significant increases in chloride in soils increase crop cadmium uptake. In the long term, cadmium uptake will remain high in saline soils without changes in soil salinity. Natural soil salinity is difficult to ameliorate. Salinity of irrigation waters may be altered in some circumstances, but such modification is likely to be uneconomic.

Soil type, tillage and organic matter

Characteristics include soil texture, soil mineralogy, cation exchange capacity, organic matter, rooting volume, etc. These characteristics will affect both the amount and availability of soil cadmium. At the same concentration of soil cadmium, soil texture has a large effect on cadmium uptake. Soil texture cannot be changed easily.

In the long term, soils with higher organic matter content have lower soil-plant transfer of added cadmium. Increases in soil organic matter in the short term will reduce cadmium uptake by crops.

Tillage practices could perhaps affect cadmium uptake in the long term; for example, minimum tillage could cause stratification of cadmium, pH, organic matter, etc., which in relation to rooting pattern in different soil horizons will affect uptake of cadmium.

Additives

Possible soil additives to reduce cadmium uptake include zeolites, industrial clays, iron and manganese that become hydrous oxides, iron and manganese-rich (low cadmium) sewage sludges, or composts. There is some laboratory evidence to suggest that cadmium uptake can be reduced by addition of such materials in the short term. There are no field data to confirm whether these additives are cost effective in the short or long term. Any additive needs to undergo risk assessment to assess potential adverse environmental effects.

Climate

Root distribution with depth will be affected by the variations in soil water content. Proliferation of roots in moist cadmium-enriched surface soil will increase cadmium uptake. This could help to explain large year-to-year variations in cadmium concentrations in crops. However, rainfall cannot be changed and irrigation is not always available.

Interactions between factors

Interactions between the above factors may be important. For example, additions of liming materials or phosphate may induce zinc deficiency and lead to greater cadmium uptake by crops. Such interactions may be of economic importance, but are little understood.

Prioritisation of management options

Overall, cadmium inputs to the soil should be minimised.

Options for cadmium management vary regionally due to differences in environmental and soil factors, and this may lead to different priorities for applying the cadmium management options listed above. Assessing and implementing different management practices leads to a prescription approach applicable for use on local and regional scales.

Using such an approach, it is assumed that good nitrogen and phosphate fertilizer management is practised in all agricultural systems in terms of inputs, timing and placement.

Examples of a prescription approach are:

- In regions with moist acid soils, initial attention would be paid to the use of lime, zinc, crop species/crop use (e.g. animal feed crops vs. food use), and appropriate cultivars.

- In dry regions with acid soils, the same factors would also be important.

- With neutral soils, crop/crop use, cultivars and zinc would be the first considerations to control cadmium uptake.

Longevity, environmental and other implications of control measures

In most agricultural systems, good practice seeks to sustain agricultural productivity with minimum adverse environmental impact. Most of the management factors identified above constitute good agricultural practice. However, excessive use of zinc could cause phytotoxicity problems and lime-induced manganese deficiency may occur in some soils when liming materials are applied to control cadmium phytoavailability.

Some of the factors which were noted to reduce crop cadmium concentrations may also reduce crop cadmium bioavailability. Higher soil zinc and crop zinc have reduced cadmium absorption by laboratory animals, but we are not aware of any human studies of the potential effect of food zinc on absorption of food cadmium. Similar human cadmium bioavailability experiments are needed to evaluate such potential benefits from soil treatments.

LIST OF PARTICIPANTS
IN THE OECD CADMIUM WORKSHOP

Name	Address	Tel/(Fax)	[1] S	[2] F	[3] WG
AUSTRALIA					
Badenoch-Jones, Dr. Jane	Environmental Health + Safety Unit Department of Human Services and Health GPO Box 9848 Canberra ACT 2601 AUSTRALIA	+61-6-289-7110 f. +61-6-289-7222		X	X
Borrowman, Hugh	Australian Embassy Sergels Torg 12 Box 7003 103 86 Stockholm SWEDEN	+46-8-613 29 00 f. +46-8-24 74 14	X	X	X
Crighton, John	Multilateral Trade Organisations Branch Dept. of Foreign Affairs & Trade Canberra ACT AUSTRALIA	+61-6-261 2777 f. +61-6-273 1527	X	X	X
Evans, Ray	Executive Officer Western Mining Corporation 1 Southbank Boulevard South Melbourne, Vic AUSTRALIA	+61-3-685 60 00 f. +61-3-685 64 00 f. +61-3-686 35 69	X	X	X
Harrison, Stanford	Department of Primary Industries & Energy Agricultural & Veterinary Chemicals Section, GPO Box 858 Canberra ACT 2601 AUSTRALIA	+61-6-272 5405 f. +61-6-272 5899		X	X
Hughes, David	MIM Holdings Limited GPO Box 1433 410 Ann Street Brisbane QLD 4000 AUSTRALIA	+61-7-3833 8056 f. +61-7-3833 8442	X	X	X
Hunt, Brian G.	Executive Director Fertilizer Ind. Federation of Australia GPO Box 2134 T Melbourne VIC 3001 AUSTRALIA	+61-3-602 36 88 f. +61-3-602 39 77	X	X	X
McLaughlin, Dr. Michael J.	CSIRO Division of Soils PMB 2 Private Bag No. 2 Glen Osmond SA 5004 AUSTRALIA	+61-8-303 8433 f. +61-8-303 8565	X	X	X
Spencer, Dr Terry	Principal Food Technologist National Food Authority PO Box 7186 Canberra MC, ACT 2610 AUSTRALIA	+61-6-271 22 86 f. +61-6-271 22 09	X	X	X
Williams, Dr. Richard	Principal Research Scientist Agricultural Production and Natural Resources Branch Bureau of Resource Science PO Box E11 Queen Victoria Terrace ACT 2600 AUSTRALIA	+61-6-272 5197 f. +61-6-272 4896	X	X	X

[1] Sources Workshop [2] Fertilizer Workshop [3] The Working Group of Policy Experts

BELGIUM					
Scokart, Paul	Institute for Chemical Research Ministry of Agriculture Leuvensesteenweg 17 B-3080 Tervuren BELGIUM	+32-2-767 53 01 f. +32-2-767 72 88		X	X
CANADA					
Bailey, Dr. Loraine	Agriculture and Agri-Food Canada Research Centre P.O. Box 1000A, RR #3 Brandon, Manatoba R7A 5Y3, CANADA	+1-204-726-7650 f. +1-204-728-3858	X	X	
Bordin, Dennis	Falconbridge Ltd. Kidd Creek Division 663 Richelieu Street Custom Materials, P.O. Box 2002 Timmins, Ontario, P4N 5G5 CANADA	+1-705-235 7776 f. +1-705-235 7433	X	X	(X)
Buccini, John	Environment Canada Place Vincent Massey, 14th floor Hull, Quebec K1A 0H3 CANADA	+1-819-997-1499 f. +1-819-953-4936			X
Chandler, John	A.J. Chandler & Associates Ltd. Environmental Management Consultants 12 Urbandale Avenue, Willowdale Ontario M2M 2H1 CANADA	+1-416-250 6570 f. +1-416-733 2588	X	X	
Clapham, Michael	Industry Spec. Environment & Technology Metals & Minerals Directorate Resource Processing Ind. Branch 235 Queen Street, 9th floor, East Tower (955B) Ottawa, Ontario K1A 0145 CANADA	+1-613-954 3123 f. +1-613-954 3079			X
Finlay, Patrick	Environment Canda Mining, Minerals & Metals Division Environmental Protection Service Place Vincent Massey, 13th floor 351 St. Joseph Blvd. Hull, Quebec K1A 0H3 CANADA	+1-819-953 1103 f. +1-819-953 5053	X	X	
Garrett, Robert	Geological Survey of Canada 601 Booth St. Ottawa, Ontario K1A 0E8 CANADA	+1-613-995-4517 f. +1-613-996-3726	X	X	
Grant, Dr. Cindy	Agriculture Canada Agriculture and Agri-Food Canada Research Centre P.O. Box 1000A Brandon, Manatoba R7A 5Y3 CANADA	+1-204-726-7650 f. +1-204-728-3858	X	X	

Hill, Robin	Health and Welfare Canada Environmental Health Centre Tunney's Pasture Ottawa, Ontario K1A OL2 CANADA	+1-613-941-3993 f. +1-613-941-4546	X	X	X
Keating, John	Natural Resources Canada Commodity Specialist Nonferrous Division 580 Booth Street Ottawa, Ontario K1A 0E4 CANADA	+1-613-992 44 09 f. +1-613-943 84 50	X	X	X
Kenny, Margaret	Agriculture Canada FP&I Branch - Fertilizer Section 59 Camelot - 1st Floor Nepean Ontario, K1A 0Y9 CANADA	+1-613-952-8000 f. +1-613-992-5219	X	X	X
Koren, Elaine	Foreign Affairs and International Trade Senior Policy Advisor Lester B. Pearson Building Tower C. 6th floor 125 Sussex Drive Ottawa, Ontario K1A 0G2 CANADA	+1-613-992-7883 f. +1-613-944 0064	X	X	(X)
Mudroch, Alena	Aquatic Ecosystem Restoration Branch National Water Research Institute Canada Centre for Inland Waters Box 5050 867 Lakeshore Road, Burlington Ontario L7R 4A6 CANADA	+1-905-336 4707 f. +1-905-336 6430	X	X	
Roberts, Dr. Terry	Potash and Phosphate Inst. Suite 704, CN Tower Midtown Plaza Saskatoon, Saskatchewan S7K 1J5 CANADA	+1-306-652-3535 f. +1-306-664-8941	X	X	(X)
DENMARK					
Busch, Niels Juul	Ramboll Bredevej 2 DK-2830, Virum DENMARK	+45-45-98 83 00 f. +45-45-98 85 20	X		
Heron, Henri	Ministry of Environment and Energy, Danish Environmental Protection Agency Strandgade 29 DK-1401 Copenhagen K DENMARK	+45-32-66 01 00 f. +45-32-66 04 79	X		X
Tjell, Jens Christian	Inst. of Environmental Science and Engineering Technical Univ. of Denmark Build. 115 DK-2800 Lyngby DENMARK	+45-45-25 1617 f. +45-45-93 28 50	X	X	X

FINLAND					
Bergholm, Kari	Ministry for Foreign Affairs P.O. Box 176 FIN-00161 Helsinki FINLAND	+358-0-134 15 909 +358-0-134 16 477			X
Hero, Heikki	Kemira Agro P.O. Box 330 Fin-00101 Helsinki FINLAND	+358-0-132 11 f. +358-0-1321 619		X	
Louekari, Kimmo	Finnish Inst. of Occupational Health Topeliuksenkatu 41 a A Fin-00250 Helsinki FINLAND	+358-0-4747 617 f. +358-0-4747 208	X	X	
Mäkelä-Kurtto, Ritva	Agricultural Research Center Institute of Soils and Environment Fin-31600 Jokioinen FINLAND	+358-16-4188 388 f. +358-16-4188 396	X	X	
Nikunen, Esa	Finnish Environment Agency P.O.Box 140 Fin-00251 Helsinki FINLAND	+358-0-40 300 536 f. +358-0-40 300 591	X		
FRANCE					
Delmas, A.	INRA Station des Sciences du Sol 65, rue de Saint Brieuc 35042 RENNES CEDEX FRANCE	+33-99-28 54 20 f. +33-99-28 54 30	X	X	X
Diderich, Bob	Ministère de l'Environnement DPPR/SDPD/BSPC 20, av. de Ségur 75302 Paris 07 SP FRANCE	+33-1-42 19 15 44 f. +33-1-42 19 14 68	X	X	X
Fitoussi, Richard	Rhone-Poulenc Rare Earths & Gallium 25 Quai Paul Dovmer 92408 Courbevoie Cedex FRANCE	+33-1-47 68 10 74 f. +33-1-47 68 09 99	X		
Samec, Dr Francois	Grande Paroisse S.A. Immeuble Iris 12 Place de l'Iris 92062 Paris-la-Defense Cedex 54 FRANCE	+33-1-47 96 95 85 f. +33-1-47 96 95 76	X	X	
GERMANY					
Schmid, Dr. Elisabeth	Umweltsbundesamt. Bismarck Platz 1 14193 Berlin GERMANY	+49-308-903 2615 +49-308-903 22 85	X	X	X
Wunderlich, Dietmar	Umweltbundesamt Postfach 330022 14191 Berlin GERMANY	+49-890 322 64 f. +49-890 322 85		X	

Name	Address	Phone/Fax			
Lithner, Göran	Institute of Environmental Research, Laboratory for Aquatic Environmental Chemistry, ITM Solna, Sthlm Univ. S-106 91 Stockholm SWEDEN	+46-8-674 70 00 f. +46-8-28 48 29	X		
Meijersjö, Else-Marie	Agricultural Board Vallgatan 8 551 82 Jönköping SWEDEN	+46-36-15 50 00 f. +46-36-71 05 17		X	
Nordberg, Gunnar	University of Umeå Inst. för miljömedicin 901 87 Umeå SWEDEN	+46-90-10 27 27 (+46-90-10 17 00) f. +46-90-77 96 30)	X	X	
Noreus, Dag	Department of Structural Chemistry, Stockholm University S-106 91 Stockholm SWEDEN	+46-8-16 12 53 f. +46-8-15 21 87	X		
Olsson, Sigrid	Environmental Protection Agency Blekholmsterrassen 36 106 48 STOCKHOLM SWEDEN	+46-8-698 10 00 f. +46-8-20 29 25	X		
Olsson-Öberg, Mona	ÅF-Industrins Processkonsult Box 47086 402 58 Göteborg SWEDEN	+46-31-46 00 80 f. +46-31-48 21 80		X	
Petersson-Grawe, Kierstin	National Food Board Hamnesplanaden 5 Box 622 751 26 Uppsala SWEDEN	+46-18-17 55 00 f. +46-18-10 58 48		X	
Stråby, Arne	Occupational Safety and Health Board Ekelundsvägen 16 171 84 SOLNA SWEDEN	+46-8-730 90 00 f. +46-8-730 19 67	X		
Turesson, Anders	Ministry of Environment Tegelbacken 2 103 33 STOCKHOLM SWEDEN	+46-8-405 10 00 f. +46-8-21 91 70	X	X	
Wahlström, Bo - chairman	National Chemicals Inspectorate Sundbybergsvägen 9 Box 1384 171 27 SOLNA SWEDEN	+46-8-730 6717 f. +46-8-735 76 98	X	X	X
Öberg, Karin	Environment Protection Agency Blekholmsterrassen 36 106 48 STOCKHOLM SWEDEN	+46-8-698 10 00 f. +46-8-698 13 45	X		
Öborn, Ingrid	Agriculture University of Uppsala Ullsväg 17 Box 7014 750 07 Uppsala SWEDEN	+46-18- 67 10 00 f. +46-18-30 15 53		X	
SWITZERLAND					
Dettwiler, Johannes	Bundesamt für Umwelt, Wald und Landschaft (Buwal) Hallwylstrasse 4 CH-3003 Bern SWITZERLAND	+41-31-322 93 46 f. +41-31-324-79 78	X	X	X

JAPAN

Name	Address	Phone/Fax			
Fujimoto, Kinya	Japan Storage Battery Ass. 3-5-8 Shibakouen Minato-Ku, Tokyo ID5 JAPAN	+81-3-3434 0261 f. 81-3-3434 2691	X		
Kasuga, Kenji	Deputy Director Chemical Fertilizer Office Basic Industries Bureau Ministry of International Trade and Industry 1-3-1 Kasumigaseki Chiyoda-Ku Tokyo JAPAN	+81-3-3501-1707 f. +81-3-3501 4721		X	
Kawamata, Motoo	Japan Chemical Industry Ass. Tokyo Club Bldg 2-6 3-chome, Kasumigaseki Chiyoda-Ku, Tokyo JAPAN	+81-3-3580-1381 f. +81-3-3580-1383	X	X	X
Mukunoki, Junij	Japan Storage Battery Ass. 3-5-8 Shibakouen Minato-Ku, Tokyo ID5 JAPAN	+81-3-3434 0261 f. 81-3-3434 2691	X		
Sasakura, Jun	Toshiba Battery Co, Ltd Sumitomo-Fudosan Shinagawa 10-27, Higashi-Shinaga Wa, 4-chome Shinagawa-ku Tokyo JAPAN	+81-3-5460-7560 f. +81-3-5460 5323	X		
Suzurida, Kazuhiro	Import Team Fertilizer Department Mitsubishi Corporation 2-3-1 Marunouchi Chiyoda-Ku Tokyo JAPAN	+81-3-3210 8343 f. +81-3-3210 8337		X	

NETHERLANDS

Name	Address	Phone/Fax			
Coster, Rein	Vereninging van Kunstmest Producenten P.O. Box 443 2260 AK Leidschendam The NETHERLANDS	+31-70-337 87 18 f. +31-70-320 39 03		X	
del Castilho, Pierre	Ministry of Agriculture, Nature management and Fisheries AB-DLO, P.O. Box 129 9750 AC Haren The NETHERLANDS	+31-50-33 75 07 f. +31-50-33 73 93		X	
Dorgelo, Folke	Ministry of Housing Chemicals Division/ 655 8, Rijnstraat, P O Box 30945 2500 GX, The Hague The NETHERLANDS	+31-70-339 4908 f. +31-70-339 1297	X	X	X
Eijsackers, Herman	RIVM. P.O. Box 1 NL-3720 BA Bilthoven Bilthoven The NETHERLANDS	+31-30-74 30 15 f. +31-30-25 19 25	X		

Hansen, Erik	Cowiconsult A/S Flegborg 6 DK-7100 VEJLE DENMARK	+45-76-42 64 24 f. +45-76-42 64 01 (f.+45-75836273)	X		
Koot, J.E.	N.F.K. (Netherlands Federation for Plastics) Vtietweg 16 Postbox 420 NL-2260 AK Leidschendam The NETHERLANDS	+31-70-317 54 90 f. +31-70-317 74 08	X		
Moolenaar, Simon	Wageningen Agricultural University Department of Soil Science and Plant Nutrition P.O. Box 8005 6700 EC Wageningen The NETHERLANDS	+31-317-48 23 50 f. +31-317-48 37 66	X	X	
Pearse, J.D.	Ballard Point House 24 Burlington Road Swanage BH19, ILT UNITED KINGDOM	+44-1-929-422 700 f. +44-1-929-422 700	X	X	X
Sedee, Dr. Aad.G.J.	Ministry of Housing, Physical Planning and Environment Directorate for Chemicals, External Safety and Radiation Protection, ipc 655, Rijnstraat 8 P O Box 30945 2500 GX, The Hague The NETHERLANDS	+31-70-339 49 44 f. +31-70-339 1297	X	X	X
Sundberg, Viktor	Electrolux Avenue de Tervuereu 12/bte 3 B-1040 BRUSSELS BELGIUM	+32-2-735 09 30 f. +32-2-732 33 99	X		
Vermeul, Robert	Hydro Agri Rotterdam B.V. afdeling Development Maassluissedijk 103 3130 AB Vlaardingen The NETHERLANDS	+31-10-460 74 56 f. +31-10-460 74 20		X	
Vonkeman, Prof. Dr. Gerrit H.	Institute for European Environmental Policy - Brussels Eendrachstraat 51 1050 Bruxells BELGIUM	+32-2-514 01 24 f. +32-2-512 32 65	X	X	X
NEW ZEALAND					
Furness, Dr. Hilton	New Zealand Fertilizer Manufacturers' Research Association (Inc) PO Box 9577 New Market, Auckland NEW ZEALAND	+64-9-309 9782 f. +64-9-307 6690	X	X	X
Reeve, John	MAF Regulatory Authority - Agricultural Compounds Unit PO Box 40063 Upper Hutt NEW ZEALAND	. +64-4-528 6089 f. +64-4-528 4675	X	X	X
Sinner, Jim	Ministry of Agriculture and Fisheries MAF Policy, 6th floor 101 The Terrace P.O. Box 2526 Wellington NEW ZEALAND	+64-4-474 41 00 f. +64-4-473 0118			X

NORWAY					
Høygaard, Erik	Norwegian Pollution Control Authority P.O. Box 8100 Dep. N-0032 Oslo NORWAY	+47-22-57 34 82 +47-22 57 34 00 f. +47-22-67 67 06			X
Næss, Bent Gunnar	Royal Ministry of Agriculture Department of Agricultural Affairs P.O. Box 8007, Dep. N-0030 Oslo NORWAY	+47-22-34 93 46 f. +47-22-34 95 57	X	X	
Nygreen, Beryl C.	Norwegian Pollution Control Authority P.O. Box 8100 Dep. N-0032 Oslo NORWAY	+47-22-57 36 49 +47-22 57 34 00 f. +47-22-67 67 06	X	X	X
SWEDEN					
Askling, Ulrika	Ministry of Agriculture Rosenbad 4 103 33 Stockholm SWEDEN	+46-8-405 11 95 f. +46-8-20 64 96		X	X
Bengtsson, Gunnar	Director General National Chemicals Inspectorate Sundbybergsvägen 9 Box 1384 171 27 SOLNA SWEDEN	+46-8-730 57 00 f. +46-8-735 76 98	X	X	X
Bengtsson, Leif	National Chemicals Inspectorate Sundbybergsvägen 9 Box 1384 171 27 SOLNA SWEDEN	+46-8-730 6814 f. +46-8-735 76 98	X	X	
Bergbäck, Bo	Kalmar University Dep. of Natural Science Box 905 391 29 KALMAR SWEDEN	+46-480-44 62 45 f. +46-480-44 62 62	X		
Bertilsson, Göte	Hydro Agri AB Box 516 261 24 LANDSKRONA SWEDEN	+46-418-761 67 f. +46-418 583 46		X	
Borg, Hans	Institute of Environmental Research, Laboratory for Aquatic Environmental Chemistry ITM Solna, Sthlm Univ. S-106 91 Stockholm SWEDEN	+46-8-674 70 00 f. +46-8-28 48 29	X		
Bucht, Bengt	National Chemicals Inspectorate Sundbybergsvägen 9 Box 1384 171 27 SOLNA SWEDEN	+46-8-730 6805 f. +46-8-735 76 98		X	
Dietrichson, Eva	National Chemicals Inspectorate Sundbybergsvägen 9 Box 1384 171 27 SOLNA SWEDEN	+46-8-730 68 02 f. +46-8-735 76 98	X	X	X

Name	Address	Phone/Fax			
Eksvärd, Jan	Swedish Farmers Association S:t Göransgatan 160 A Box 30192 104 25 Stockholm SWEDEN	+46-8-657 42 00 f. +46-8-618 69 12	X	X	
Eriksson, Jan	Agriculture University of Uppsala Ullsväg 17 Box 7014 750 07 Uppsala SWEDEN	+46-18- 67 10 00 f. +46-18-30 15 53		X	
Folke, Jens	Environmental Research Group Ltd Östergade 16 DK-3250 Gilleleje DANMARK	+45-48 35 45 41 f. +45-48 35 45 42		X	
Gustafsson, Lars	National Chemicals Inspectorate Sundbybergsvägen 9 Box 1384 171 27 SOLNA SWEDEN	+46-8-730 67 93 f. +46-8-735 76 98	X	X	X
Heijkenskjöld, Lolo	Ministry of Environment Batteriutredningen (M 1995:03) Tegelbacken 2 103 33 STOCKHOLM SWEDEN	+46-8-405 10 00 f. +46-8-21 91 70	X		
Iverfeldt, Åke	Swedish Environmental Research Institute Box 21060 100 31 Stockholm SWEDEN	+46-8-729 15 00 f. +46-8-31 85 16	X		
Johansson, Kjell	Environmental Protection Agency Blekholmsterrassen 36 S-106 48 Stockholm SWEDEN	+46-8-698 10 00 f. +46-8-20 29 25	X		
Järup, Lars	Karolinska Hospital 171 76 STOCKHOLM SWEDEN	+46-8-729 20 00 f.	X		
Kjällman, Anders	Environmental Protection Agency Blekholmsterrassen 36 S-106 48 Stockholm SWEDEN	+46-8-698 10 00 f. +46-8-20 29 25	X		
Klöfver, Inger	Environmental Protection Agency Blekholmsterrassen 36 106 48 STOCKHOLM SWEDEN	+46-8-698 10 00 f. +46-8-20 29 25	X		
Laanatza, Marianne	Själlandsgatan 142 164 45 Kista SWEDEN	+46-8-751 66 43 f. +46-8-751 66 84	X		
Landner, Lars	Environmental Research Group Ltd Götgatan 35 116 21 Stockholm SWEDEN	+46-8-643 37 59 f.	X	X	
Lindau, Lars	Environmental Protection Agency Blekholmsterrassen 36 106 48 STOCKHOLM SWEDEN	+46-8-698 10 00 f. +46-8-20 29 25	X		X
Lindbäck, Christina	Ministry of Environment Tegelbacken 2 103 33 STOCKHOLM SWEDEN	+46-8-405 10 00 f. +46-8-21 91 70	X		

UNITED KINGDOM					
Curtis, Mike	Department of Trade and Industry VMM 2b 151 Buckingham Palace Road London SW1W 9SS UNITED KINGDOM	+44-1-71 215 1456 f. +44-1-71 215 1070	X		X
Heather, David J.	The Fertilizer Manufacturers Association Ltd. Greenhill House Thorpe Wood Peterborough PE3 6GF UNITED KINGDOM	+44-1-733 33 1303 f. +44-1-733 33 3617		X	
Hutton, Malcolm	Environment Resource Management Ltd 8 Cavendish Square London W1M 0ER UNITED KINGDOM	+1-44 171 465 7226 f. +44-171 465 7320	X		
Knock, Bill	Ministry of Agriculture Fisheries and Food Ergon House 17 Smith Square London UNITED KINGDOM	+44-171-238 6293 +44-171-238 6338		X	X
Moxon, Richard	Department of the Environment Water Resources Marine Division Room A433, Romney House 43 Marsham St. London SW1P 3PY UNITED KINGDOM	+44-1-71 276 8268. f. +44-1-71 276 8603	X		X
Rudduck, Martin	Department of Trade and Industry CB1.126 151 Buckingham Palace Road London SW1W 9SS UNITED KINGDOM	+44-171-215 1107 f. +44-171-215 5882	X		
Simpson, Peter	British Geological Survey Nicker Hill Keyworth Nottingham NG12 5GG UNITED KINGDOM	+44-1-159 36 3532 f. +44-1-159 36 3200	X	X	X
USA					
Beal, Diane	U.S. Environmental Protection Agency, Office of Pollution prevention and toxics 401 M Street, S.W. (7401) Washington D.C. 20460 USA	+1-202-260 1822 f. +1-202-260 0575	X	X	X
Boreiko, Dr. Craig J.	Manager. Environmental Health International Lead Zinc research organization 2525 Meridian Parkway Research Triangle Park. NC 27709 USA	+1-919-361 4647 f. +1-919-361 1957	X	X	
Chaney, Rufus	US Department of Agriculture ARS Environmental Chemistry Lab Building 007, BARC-West Beltsville, MD 20705 USA	+1-301-504 8324 f. +1-301-504 5048	X	X	X

Name	Address	Phone/Fax			
England, Norman	Portable Rechargeable Battery Association 1000 Parkwood Circle - Suite 430 Atlanta. GA 30339 USA	+1-770 612 8826 f. +1-770 612 8841	X		
Gough, Dr. Larry	U.S. Geological Survey/ Cen. for Environ. Geochem. & Geophysics Denver Federal Center, MS 973 Denver CO, 80225 USA	+1-303-236 5513 f. +1-303-236 3200	X	X	X
Morrow, Hugh	International Cadmium Association 12110 Sunset Hills Road - Suite 110 Reston. VA 22090 USA	+1-703-709 1400 f. +1-703-709 1402	X	X	X
Nilsson, Arne O.	Acme Electric Corporation 528 West 21st Street Tempe Az 85282 USA	+1-602-894 6864 f. +1-602-921 0470			
Ulicny, Leonard J.	Manager of Technology SCM Chemicals 2701 Broening Highway Baltimore. MD 21222 USA	+1-410-288 8845 f. +1-410-288-8897	X		
Walker, Thomas H.	Industrial Economics, Inc. 2067 Massachusetts Avenue Cambridge MA. 02140, USA	+1-617-354 0074 f. +1-617-354 0463	X		
BIAC					
Atherton, Dr John K	International Cadmium Association European Office 42 Weymouth Street London, W1N 3LQ UNITED KINGDOM	+44-171-499 84 25 f. +44-171-486 4007	X	X	X
Cloke, Robin	G P Batteries Ltd Unit E, Wycos Road Bridgewater Somerset TA6 4BH UNITED KINGDOM	+44-1-278 445 995 f. +44-1-278 445 989	X		
Cook, Murray E	Galvanizers Association Wren's Court 56 Victoria Road Sutton Coldfield West Midlands B72 1SY UNITED KINGDOM	+44-121 355 8838 f. +44-121-355 8727	X	X	X
Davister, Armand	Quai de la Boverie, 98/091 B-4020 Liege BELGIUM	+32-41-433 661 f. +32-41-433 661		X	
Donnelly, Dr Peter	Akcros Chemicals Lankro House PO Box 1 Eccles Manchester M30 0BH UNITED KINGDOM	+44-161-785 1111 (-1253) f. +44-161-787 7518	X		
Eloy, Robert	Saft 156, Avenue de Metz F 93230 Romainville FRANCE	+33-1-49 15 36 00 f. +33-1-48 44 11 53	X		

Name	Address	Phone/Fax			
Fraser, Wayne	Environmental Director Hudson Bay Mining and Smelting Company Ltd P.O. Box 1500 Flin Flon Manitoba R8A 1N9 CANADA	+1-204-687 2171 f. +1-204-687 5793	X	X	X
Johnston, Dr. A. E.	IACR-Rothhamsted Harpenden Hertfortshire AL5 2JQ UNITED KINGDOM	+44-1582 763 133 f. +44-1582 760 981	X	X	
Kummer, Dr. Karl-Friedrich	BASF Postfach 120 D-67114 Limburgerhof GERMANY	+49-6236 68 22 45 f. +49 62 36 68 27 91		X	
Langeveld, C.P.	Amsterdam Fertilizers B.V. Fosfaatweg 48 P.O. Box 313 1013 BM Amsterdam The NETHERLANDS	+31-20-58 15 130 f. +31-20-68 68 328		X	
Persson, Björn	Hydro Agri AB Storgatan 24 Box 516 261 24 Landskrona SWEDEN	+46-418-762 71 f. +46-418-743 81		X	
Rabin, Dr. Jean-Paul	Director, Occupational Health Noranda Metallurgy, Inc 1800 McGill Avenue Suite 2400 Montreal, PQ H3A 3J6 CANADA	+1-514-982 6383 f. +1-514-982 6399	X	X	
Riley, Deidre K	Program Supervisor Environment & Public Affairs Cominco Ltd 200 Burrard Street, Suite 500 Vancouver, BC V6C 3L7 CANADA	+1-604-844 2689 +1-604-685 3019	X	X	
Steen, Ingrid	EFMA European Fertilizer Manufacturers Association 4 ave van Niuewenhuyse Box 7 1160 Brussels BELGIUM	+32-2-675 35 50 f. +32-2-675 39 61	X	X	X
Streatfield, Gordon R.	13 Canberra Crescent Meir Park, Longton Stoke-On-Trent ST3 7RA UNITED KINGDOM	+44-1782-392 617 f. n/a	X		
Van Assche, Dr Frank	European Zinc Institute 6th Floor 12, avenue de Broqueville B-1150 Brussels BELGIUM	+32-2-775 6327 f. +32-2-779 0523	X	X	
Whelan, Jack	IFA-International Fertilizer Industry Association 28 rue Marbeuf F-75008 Paris FRANCE	+33-1-42 25 27 07 f. +33-1-42 25 24 08	X	X	

European Commission

Name	Address	Phone/Fax			
Nunez, José	European Commission Directorate General III C/4 Rond Point 11, 8th level, room 3 Rue de la Loi 200 B-1049 Brussels BELGIUM	+32-2-295 81 96 f. +32-2-295 02 81		X	
Perenius, Lena	European Commission Directorate General III C/4 Rond Point 11, 8th level, room 3 Rue de la Loi 200 B-1049 Brussels BELGIUM	+32-2-295 81 96 f. +32-2-295 02 81	X	X	X

OECD

Name	Address	Phone/Fax			
Maier, Leo	OECD Directorate for Food, Agriculture and Fisheries 2, rue André-Pascal, 75775 Paris Cedex 16 FRANCE	+33-1-45 24 93 27 f. +33-1-45 24 18 90		X	X
Sigman, Richard	OECD Environmental Health och Safety Division 2, rue André-Pascal, 75775 Paris Cedex 16 FRANCE	+33-1-45 24 16 80 f. +33-1-45 24 16 75	X	X	X
van Looy, Hugo	OECD Environmental Health och Safety Division 2, rue André-Pascal, 75775 Paris Cedex 16 FRANCE	+33-42 87 27 65 f. +33-1-45 24 16 75	X	X	X

JORDAN

Name	Address	Phone/Fax			
Al-Jazi, Mamdouh	Jordan Phosphate Mines Co Sharif Al-Radi Street No 5 P O Bo 30 Amman JORDAN	+962-6-607 141 f. +962-6-682 290	X	X	X
Bashir, Dr. Saleh Suleiman	Jordan Phosphate Mines Co Sharif Al-Radi Street No 5 P O Bo 30 Amman JORDAN	+962-6-607 141 f. +962-6-682 290	X	X	X
Sinukrot, Nabil Musbah	Jordan Phosphate Mines Co Sharif Al-Radi Street No 5 P O Bo 30 Amman JORDAN	+962-6-607 141 f. +962-6-682 290		X	

MOROCCO

Name	Address	Phone/Fax			
Chik, Abdellah	Centre d'Etudes et de Recherches des Phosphates Minéraux (CERPHOS) 73-87 Bd, Moulay Ismail Roches Noires, Casablanca MOROCCO	+02-24 12 69 f. +02-24-64 41		X	

Name	Address	Phone/Fax			
Kossir, Abdelaâli	Centre d'Etudes et de Recherches des Phosphates Minéraux, (CERPHOS) 73-87 Bd, Moulay Ismail Roches Noires Casablanca MOROCCO	+02-24 12 69 f. +02-24-64 41		X	
Youzalen, El Houssain	DQE Groupe Office Cherifien des Phosphates, Angle Route d'Ei Jadida et Boulevard de la Grande Ceinture Casablanca MOROCCO	+212-2-23 00 25 f. 212-2-23 03 60		X	
NAURU					
Harris, René R.	Chairman Nauru Phosphate Corporation Nauru House, 49th level Collins Street Melbourne Victoria 3000 AUSTRALIA	+67-4-444 37 51 f. +67-4-444 37 52 +61-3-650 50 52 f. +61-3-654 74 71	X	X	X
Walker, Kenneth E.	Hon. Consul for the Republic of Nauru 17 Castlereigh Street Sydney N.S.W. 2000 AUSTRALIA	+61-2-233 80 44 f. +61-2-221 70 32	X	X	X
SENEGAL					
Kotlarevsky, Igor	Industries Chimiques du Senegal Km 18 Route de Rufisque BP/P.O.B. 3835 Dakar SENEGAL	+221-34 01 22 f. +221-34 07 01		X	
TOGO					
Desanti, Jean-Gerard	Office Togolais des Phosphates 23 rue Francois 1er 75008 Paris FRANCE	+33-1-47 20 98 88 f. +33-1-47 20 13 88	X	X	
Ekue, Adama	Office Togolais des Phosphates 23 rue Francois 1er 75008 Paris FRANCE	+33-1-47 20 98 88 f. +33-1-47 20 13 88	X	X	
IMPHOS (World Phosphate Institute)					
Belmehdi, Abdelatif	World Phosphate Institute (IMPHOS) - OCP Building, Angle Route d'Ei Jadida et Boulevard de la Grande Ceinture Casablanca MOROCCO	+212-2-23 00 25 (ext. 38-02) f. +212-2-23 06 40		X	
Mrabet, Tayeb	World Phosphate Institute (IMPHOS)- OCP Building, Angle Route d'Ei Jadida et Boulevard de la Grande Ceinture Casablanca MOROCCO	+212-2-23 00 25 (ext. 38-02) f. +212-2-23 06 40		X	

Workshop Secretariat					
Birberg, Winnie	National Chemicals Inspectorate Sundbybergsvägen 9 Box 1384 171 27 SOLNA SWEDEN	+46-8-730 65 26 f. +46-8-735 76 98			
Helm, Kerstin	National Chemicals Inspectorate Sundbybergsvägen 9 Box 1384 171 27 SOLNA SWEDEN	+46-8-730 67 23 f. +46-8-735 76 98			
Joandi, Linda	National Chemicals Inspectorate Sundbybergsvägen 9 Box 1384 171 27 SOLNA SWEDEN	+46-8-730 68 01 f. +46-8-735 76 98			
Linder, Ulla	National Chemicals Inspectorate Sundbybergsvägen 9 Box 1384 171 27 SOLNA SWEDEN	+46-8-730 67 45 f. +46-8-735 76 98			
Wistrand, Per-Håkan *Responsible for the computer set-up*	National Chemicals Inspectorate Sundbybergsvägen 9 Box 1384 171 27 SOLNA SWEDEN	+46-8-730 68 13 f. +46-8-735 76 98			

MAIN SALES OUTLETS OF OECD PUBLICATIONS
PRINCIPAUX POINTS DE VENTE DES PUBLICATIONS DE L'OCDE

AUSTRALIA – AUSTRALIE
D.A. Information Services
648 Whitehorse Road, P.O.B 163
Mitcham, Victoria 3132   Tel. (03) 9210.7777
                              Fax: (03) 9210.7788

AUSTRIA – AUTRICHE
Gerold & Co.
Graben 31
Wien I   Tel. (0222) 533.50.14
                              Fax: (0222) 512.47.31.29

BELGIUM – BELGIQUE
Jean De Lannoy
Avenue du Roi, Koningslaan 202
B-1060 Bruxelles   Tel. (02) 538.51.69/538.08.41
                              Fax: (02) 538.08.41

CANADA
Renouf Publishing Company Ltd.
1294 Algoma Road
Ottawa, ON K1B 3W8   Tel. (613) 741.4333
                              Fax: (613) 741.5439
Stores:
61 Sparks Street
Ottawa, ON K1P 5R1   Tel. (613) 238.8985

12 Adelaide Street West
Toronto, ON M5H 1L6   Tel. (416) 363.3171
                              Fax: (416)363.59.63

Les Éditions La Liberté Inc.
3020 Chemin Sainte-Foy
Sainte-Foy, PQ G1X 3V6   Tel. (418) 658.3763
                              Fax: (418) 658.3763

Federal Publications Inc.
165 University Avenue, Suite 701
Toronto, ON M5H 3B8   Tel. (416) 860.1611
                              Fax: (416) 860.1608

Les Publications Fédérales
1185 Université
Montréal, QC H3B 3A7   Tel. (514) 954.1633
                              Fax: (514) 954.1635

CHINA – CHINE
China National Publications Import
Export Corporation (CNPIEC)
16 Gongti E. Road, Chaoyang District
P.O. Box 88 or 50
Beijing 100704 PR   Tel. (01) 506.6688
                              Fax: (01) 506.3101

CHINESE TAIPEI – TAIPEI CHINOIS
Good Faith Worldwide Int'l. Co. Ltd.
9th Floor, No. 118, Sec. 2
Chung Hsiao E. Road
Taipei   Tel. (02) 391.7396/391.7397
                              Fax: (02) 394.9176

CZECH REPUBLIC – RÉPUBLIQUE TCHÈQUE
National Information Centre
NIS – prodejna
Konviktská 5
Praha 1 – 113 57   Tel. (02) 24.23.09.07
                              Fax: (02) 24.22.94.33
(*Contact* Ms Jana Pospisilova, nkposp@dec.niz.cz)

DENMARK – DANEMARK
Munksgaard Book and Subscription Service
35, Nørre Søgade, P.O. Box 2148
DK-1016 København K   Tel. (33) 12.85.70
                              Fax: (33) 12.93.87
J. H. Schultz Information A/S,
Herstedvang 12,
DK – 2620 Albertslung   Tel. 43 63 23 00
                              Fax: 43 63 19 69
Internet: s-info@inet.uni-c.dk

EGYPT – ÉGYPTE
The Middle East Observer
41 Sherif Street
Cairo   Tel. 392.6919
                              Fax: 360-6804

FINLAND – FINLANDE
Akateeminen Kirjakauppa
Keskuskatu 1, P.O. Box 128
00100 Helsinki

Subscription Services/Agence d'abonnements :
P.O. Box 23
00371 Helsinki   Tel. (358 0) 121 4416
                              Fax: (358 0) 121.4450

FRANCE
OECD/OCDE
Mail Orders/Commandes par correspondance :
2, rue André-Pascal
75775 Paris Cedex 16   Tel. (33-1) 45.24.82.00
                        Fax: (33-1) 49.10.42.76
                        Telex: 640048 OCDE
Internet: Compte.PUBSINQ@oecd.org
Orders via Minitel, France only/
Commandes par Minitel, France exclusivement :
36 15 OCDE

OECD Bookshop/Librairie de l'OCDE :
33, rue Octave-Feuillet
75016 Paris   Tél. (33-1) 45.24.81.81
                             (33-1) 45.24.81.67

Dawson
B.P. 40
91121 Palaiseau Cedex   Tel. 69.10.47.00
                              Fax: 64.54.83.26

Documentation Française
29, quai Voltaire
75007 Paris   Tel. 40.15.70.00

Economica
49, rue Héricart
75015 Paris   Tel. 45.75.05.67
                              Fax: 40.58.15.70

Gibert Jeune (Droit-Économie)
6, place Saint-Michel
75006 Paris   Tel. 43.25.91.19

Librairie du Commerce International
10, avenue d'Iéna
75016 Paris   Tel. 40.73.34.60

Librairie Dunod
Université Paris-Dauphine
Place du Maréchal-de Lattre de Tassigny
75016 Paris   Tel. 44.05.40.13

Librairie Lavoisier
11, rue Lavoisier
75008 Paris   Tel. 42.65.39.95

Librairie des Sciences Politiques
30, rue Saint-Guillaume
75007 Paris   Tel. 45.48.36.02

P.U.F.
49, boulevard Saint-Michel
75005 Paris   Tel. 43.25.83.40

Librairie de l'Université
12a, rue Nazareth
13100 Aix-en-Provence   Tel. (16) 42.26.18.08

Documentation Française
165, rue Garibaldi
69003 Lyon   Tel. (16) 78.63.32.23

Librairie Decitre
29, place Bellecour
69002 Lyon   Tel. (16) 72.40.54.54

Librairie Sauramps
Le Triangle
34967 Montpellier Cedex 2 Tel. (16) 67.58.85.15
                           Fax: (16) 67.58.27.36

A la Sorbonne Actual
23, rue de l'Hôtel-des-Postes
06000 Nice   Tel. (16) 93.13.77.75
                              Fax: (16) 93.80.75.69

GERMANY – ALLEMAGNE
OECD Bonn Centre
August-Bebel-Allee 6
D-53175 Bonn   Tel. (0228) 959.120
                              Fax: (0228) 959.12.17

GREECE – GRÈCE
Librairie Kauffmann
Stadiou 28
10564 Athens   Tel. (01) 32.55.321
                              Fax: (01) 32.30.320

HONG-KONG
Swindon Book Co. Ltd.
Astoria Bldg. 3F
34 Ashley Road, Tsimshatsui
Kowloon, Hong Kong   Tel. 2376.2062
                              Fax: 2376.0685

HUNGARY – HONGRIE
Euro Info Service
Margitsziget, Európa Ház
1138 Budapest   Tel. (1) 111.62.16
                              Fax: (1) 111.60.61

ICELAND – ISLANDE
Mál Mog Menning
Laugavegi 18, Pósthólf 392
121 Reykjavik   Tel. (1) 552.4240
                              Fax: (1) 562.3523

INDIA – INDE
Oxford Book and Stationery Co.
Scindia House
New Delhi 110001   Tel. (11) 331.5896/5308
                              Fax: (11) 371.8275
17 Park Street
Calcutta 700016   Tel. 240832

INDONESIA – INDONÉSIE
Pdii-Lipi
P.O. Box 4298
Jakarta 12042   Tel. (21) 573.34.67
                              Fax: (21) 573.34.67

IRELAND – IRLANDE
Government Supplies Agency
Publications Section
4/5 Harcourt Road
Dublin 2   Tel. 661.31.11
                              Fax: 475.27.60

ISRAEL – ISRAËL
Praedicta
5 Shatner Street
P.O. Box 34030
Jerusalem 91430   Tel. (2) 52.84.90/1/2
                              Fax: (2) 52.84.93

R.O.Y. International
P.O. Box 13056
Tel Aviv 61130   Tel. (3) 546 1423
                              Fax: (3) 546 1442

Palestinian Authority/Middle East:
INDEX Information Services
P.O.B. 19502
Jerusalem   Tel. (2) 27.12.19
                              Fax: (2) 27.16.34

ITALY – ITALIE
Libreria Commissionaria Sansoni
Via Duca di Calabria 1/1
50125 Firenze   Tel. (055) 64.54.15
                              Fax: (055) 64.12.57
Via Bartolini 29
20155 Milano   Tel. (02) 36.50.83

Editrice e Libreria Herder
Piazza Montecitorio 120
00186 Roma   Tel. 679.46.28
                                    Fax: 678.47.51

Libreria Hoepli
Via Hoepli 5
20121 Milano   Tel. (02) 86.54.46
                                    Fax: (02) 805.28.86

Libreria Scientifica
Dott. Lucio de Biasio 'Aeiou'
Via Coronelli, 6
20146 Milano   Tel. (02) 48.95.45.52
                                    Fax: (02) 48.95.45.48

JAPAN – JAPON
OECD Tokyo Centre
Landic Akasaka Building
2-3-4 Akasaka, Minato-ku
Tokyo 107   Tel. (81.3) 3586.2016
                                    Fax: (81.3) 3584.7929

KOREA – CORÉE
Kyobo Book Centre Co. Ltd.
P.O. Box 1658, Kwang Hwa Moon
Seoul   Tel. 730.78.91
                                    Fax: 735.00.30

MALAYSIA – MALAISIE
University of Malaya Bookshop
University of Malaya
P.O. Box 1127, Jalan Pantai Baru
59700 Kuala Lumpur
Malaysia   Tel. 756.5000/756.5425
                                    Fax: 756.3246

MEXICO – MEXIQUE
OECD Mexico Centre
Edificio INFOTEC
Av. San Fernando no. 37
Col. Toriello Guerra
Tlalpan C.P. 14050
Mexico D.F.   Tel. (525) 665 47 99
                                    Fax: (525) 606 13 07

NETHERLANDS – PAYS-BAS
SDU Uitgeverij Plantijnstraat
Externe Fondsen
Postbus 20014
2500 EA's-Gravenhage   Tel. (070) 37.89.880
Voor bestellingen: Fax: (070) 34.75.778

Subscription Agency/
Agence d'abonnements :
SWETS & ZEITLINGER BV
Heereweg 347B
P.O. Box 830
2160 SZ Lisse   Tel. 252.435.111
                                    Fax: 252.415.888

**NEW ZEALAND –
NOUVELLE-ZÉLANDE**
GPLegislation Services
P.O. Box 12418
Thorndon, Wellington   Tel. (04) 496.5655
                                    Fax: (04) 496.5698

NORWAY – NORVÈGE
NIC INFO A/S
Ostensjoveien 18
P.O. Box 6512 Etterstad
0606 Oslo   Tel. (22) 97.45.00
                                    Fax: (22) 97.45.45

PAKISTAN
Mirza Book Agency
65 Shahrah Quaid-E-Azam
Lahore 54000   Tel. (42) 735.36.01
                                    Fax: (42) 576.37.14

PHILIPPINE – PHILIPPINES
International Booksource Center Inc.
Rm 179/920 Cityland 10 Condo Tower 2
HV dela Costa Ext cor Valero St.
Makati Metro Manila   Tel. (632) 817 9676
                                    Fax: (632) 817 1741

POLAND – POLOGNE
Ars Polona
00-950 Warszawa
Krakowskie Prezdmiescie 7   Tel. (22) 264760
                                    Fax: (22) 265334

PORTUGAL
Livraria Portugal
Rua do Carmo 70-74
Apart. 2681
1200 Lisboa   Tel. (01) 347.49.82/5
                                    Fax: (01) 347.02.64

SINGAPORE – SINGAPOUR
Ashgate Publishing
Asia Pacific Pte. Ltd
Golden Wheel Building, 04-03
41, Kallang Pudding Road
Singapore 349316   Tel. 741.5166
                                    Fax: 742.9356

SPAIN – ESPAGNE
Mundi-Prensa Libros S.A.
Castelló 37, Apartado 1223
Madrid 28001   Tel. (91) 431.33.99
                                    Fax: (91) 575.39.98
Mundi-Prensa Barcelona
Consell de Cent No. 391
08009 – Barcelona   Tel. (93) 488.34.92
                                    Fax: (93) 487.76.59

Llibreria de la Generalitat
Palau Moja
Rambla dels Estudis, 118
08002 – Barcelona
               (Subscripcions) Tel. (93) 318.80.12
               (Publicacions) Tel. (93) 302.67.23
                                    Fax: (93) 412.18.54

SRI LANKA
Centre for Policy Research
c/o Colombo Agencies Ltd.
No. 300-304, Galle Road
Colombo 3   Tel. (1) 574240, 573551-2
                                    Fax: (1) 575394, 510711

SWEDEN – SUÈDE
CE Fritzes AB
S–106 47 Stockholm   Tel. (08) 690.90.90
                                    Fax: (08) 20.50.21

For electronic publications only/
Publications électroniques seulement
STATISTICS SWEDEN
Informationsservice
S-115 81 Stockholm   Tel. 8 783 5066
                                    Fax: 8 783 4045

Subscription Agency/Agence d'abonnements :
Wennergren-Williams Info AB
P.O. Box 1305
171 25 Solna   Tel. (08) 705.97.50
                                    Fax: (08) 27.00.71

SWITZERLAND – SUISSE
Maditec S.A. (Books and Periodicals/Livres
et périodiques)
Chemin des Palettes 4
Case postale 266
1020 Renens VD 1   Tel. (021) 635.08.65
                                    Fax: (021) 635.07.80

Librairie Payot S.A.
4, place Pépinet
CP 3212
1002 Lausanne   Tel. (021) 320.25.11
                                    Fax: (021) 320.25.14

Librairie Unilivres
6, rue de Candolle
1205 Genève   Tel. (022) 320.26.23
                                    Fax: (022) 329.73.18

Subscription Agency/Agence d'abonnements :
Dynapresse Marketing S.A.
38, avenue Vibert
1227 Carouge   Tel. (022) 308.08.70
                                    Fax: (022) 308.07.99

See also – Voir aussi :
OECD Bonn Centre
August-Bebel-Allee 6
D-53175 Bonn (Germany)   Tel. (0228) 959.120
                                    Fax: (0228) 959.12.17

THAILAND – THAÏLANDE
Suksit Siam Co. Ltd.
113, 115 Fuang Nakhon Rd.
Opp. Wat Rajbopith
Bangkok 10200   Tel. (662) 225.9531/2
                                    Fax: (662) 222.5188

**TRINIDAD & TOBAGO, CARIBBEAN
TRINITÉ-ET-TOBAGO, CARAÏBES**
SSL Systematics Studies Limited
9 Watts Street
Curepe
Trinadad & Tobago, W.I.   Tel. (1809) 645.3475
                                    Fax: (1809) 662.5654

TUNISIA – TUNISIE
Grande Librairie Spécialisée
Fendri Ali
Avenue Haffouz Imm El-Intilaka
Bloc B 1 Sfax 3000   Tel. (216-4) 296 855
                                    Fax: (216-4) 298.270

TURKEY – TURQUIE
Kültür Yayinlari Is-Türk Ltd. Sti.
Atatürk Bulvari No. 191/Kat 13
06684 Kavaklidere/Ankara
                              Tél. (312) 428.11.40 Ext. 2458
                                    Fax : (312) 417.24.90
                                    et 425.07.50-51-52-53

Dolmabahce Cad. No. 29
Besiktas/Istanbul   Tel. (212) 260 7188

UNITED KINGDOM – ROYAUME-UNI
HMSO
Gen. enquiries   Tel. (0171) 873 0011
Postal orders only:
P.O. Box 276, London SW8 5DT
Personal Callers HMSO Bookshop
49 High Holborn, London WC1V 6HB
                                    Fax: (0171) 873 8463
Branches at: Belfast, Birmingham, Bristol,
Edinburgh, Manchester

UNITED STATES – ÉTATS-UNIS
OECD Washington Center
2001 L Street N.W., Suite 650
Washington, D.C. 20036-4922 Tel. (202) 785.6323
                                    Fax: (202) 785.0350
Internet: washcont@oecd.org
Subscriptions to OECD periodicals may also be
placed through main subscription agencies.

Les abonnements aux publications périodiques de
l'OCDE peuvent être souscrits auprès des
principales agences d'abonnement.

Orders and inquiries from countries where Distributors have not yet been appointed should be sent to:
OECD Publications, 2, rue André-Pascal, 75775
Paris Cedex 16, France.

Les commandes provenant de pays où l'OCDE n'a
pas encore désigné de distributeur peuvent être
adressées aux Éditions de l'OCDE, 2, rue André-
Pascal, 75775 Paris Cedex 16, France.

8-1996

OECD PUBLICATIONS, 2, rue André-Pascal, 75775 PARIS CEDEX 16
PRINTED IN FRANCE
(97 96 14 1) ISBN 92-64-15342-X – No. 49131 1996